Lecture Notes in Economics and Mathematical Systems

Managing Editors: M. Beckmann and W. Krelle

Economic Theory

217

Christiaan Grootaert

The Relation Between Final Demand and Income Distribution

With Application to Japan

Springer-Verlag Berlin Heidelberg GmbH

Author

Dr. Christiaan Grootaert
Economics Department, University of Hawaii at Manoa
Honolulu, HI 96822, USA

ISBN 978-3-540-12307-1   ISBN 978-3-642-51563-7 (eBook)
DOI 10.1007/978-3-642-51563-7

Library of Congress Cataloging in Publication Data. Grootaert, Christiaan, 1950-
The relation between final demand and income distribution, with application to Japan. (Lecture notes in economics and mathematical systems; 217) Bibliography: p.
1. Income distribution–Japan–Mathematical models. 2. Japan–Economic conditions–1945–Mathematical models. I. Title. II. Series. HC465.I5C76 1983 339.2'0952 83-12443

Originally published by Springer-Verlag Berlin Heidelberg New York Tokyo in 1983

2142/3140-543210

TO MY PARENTS,

with love and respect.

ACKNOWLEDGEMENTS

To a large extent this paper is based on work done at the University of Hawaii at Manoa. I would like to express my gratitude to Professors Seiji Naya, George Akita, Burnham Campbell, Moheb Ghali, and Youngil Lim, for their continuous encouragement and patience in reviewing earlier drafts and in guiding me through the work. I am also indebted to Professor Mizoguchi of Hitotsubashi University for many helpful comments. Later research incorporated in the present text benefitted from valuable input from Jan Bisschop and Graham Pyatt. Useful suggestions for revisions leading to the present manuscript were provided by the editor of the series Lecture Notes in Economics and Mathematical Systems and by an unknown referee.

ABSTRACT

This paper investigates the impact on income distribution of changes in the structure of final demand that occur in the process of economic growth.

The model used for the analysis is an equilibrium model that generates both the level and distribution of household income, and which has as key features the sectoral breakdown of each final demand component, the explicit recognition of differences in spending patterns between socio-economic groups, and the inclusion of feedback effects between the processes of spending, production, and income generation.

Methodologically, this study belongs to the growing volume of literature on the relationship between income distribution and economic growth. One of the more recent lines of research in this field is the attempt to integrate theories of distribution and growth by investigating economic interdependence between various groups of households in society. This study follows in this path since the core element of the solution of the model is a matrix describing demand linkages between income groups. This matrix indicates lower bound income inequality levels that are compatible with some structural characteristics of the economy captured by the model, both in short-run equilibrium and in a long-run steady state situation. The comparison of these two levels with the observed levels of inequality yields new insights into distributional trends not derivable from other methods of analysis. It also has important implications for the selection of redistributive policy measures.

An application of the model is presented for Japan in the period 1959-1969, a decade during which the Japanese economy grew rapidly and was not disturbed by any major external shocks. A solution and discussion of the full model is presented for each of the two years of analysis, viz. 1959 and 1969. For each year, the matrix of demand linkages between income groups and the implied minimum levels of inequality are derived.

The model solution yields the equilibrium impact of additional spending in each (input-output) sector on the income of each income group. This is then used to investigate the contribution of each component of final demand to the processes of income generation and distribution, and to establish the relationship between changes in the sectoral composition of final demand and changes in income distribution. A series of simulation experiments investigates the relative impact of changes in demand patterns, in input-output structure, and in the sectoral level and distribution of households' share of value added on the level and distribution of household income over the 1959-69 decade.

It was found that in Japan the observed increase in income inequality between 1959 and 1969 was the net effect of two opposing trends. On the one hand, demand linkages changed in favor of the middle income groups: in 1969 they derived a larger share of the income generated by the spending of other groups than in 1959. This was largely traced to changes in consumption patterns and resulted in a slight decrease in inequality. On the other hand, sectoral changes in the other final demand components, especially investment and government spending, increased income inequality exogenously, at least in the first half of the decade. In the latter part of the decade the increased income inequality resulted from changes in the level and distribution of households' share of value added, while changes in the sectoral composition of final demand and changes in the input-output structure now slowed down the increasing trend in inequality. The changing role of investment stood out in this process.

PREFACE AND OUTLINE

This paper presents a model to investigate the extent to which changes in the pattern of final demand that occur with economic growth affect income distribution. Recent research on the relationship between income distribution and economic growth has focused on integrating theories of distribution and growth by investigating economic interdependencies between various groups of households in society. This study goes along the same lines as the key element of the model is a matrix describing demand linkages between socio-economic groups. The major innovative contribution of this study is the explanation of the over-time changes in these linkages, and the tying of this explanation to the analysis of changes in income distribution, specifically to the analysis of how changes in the composition of final demand affect distribution.

Chapter I contains a brief survey of the literature dealing with the relationship between income distribution and economic growth. Emphasis is given on those models that attempt to integrate the explanation of growth and distribution. The critique of existing models will provide the background for the selection of the model used in this study.

Chapter II presents the formal derivation of the model. Each element of the solution equation is discussed in detail. Special emphasis is given to the interpretation of the matrix that describes the linkages between socio-economic groups. The assumptions underlying the model receive attention in the last section of the chapter.

Chapter III contains the application of the model to Japan during the decade 1959-1969. A review of selected studies on Japanese income distribution precedes the discussion of the data base and the empirical estimation of the parameters of the model. Complete model solutions for 1959 and 1969 are then presented and examined.

Chapter IV contains the results of a series of simulation exercises intended to discover the relative contributions of several structural changes that took place between 1959 and 1969 (including, of course, changes in the composition of final demand) to observed changes in income distribution.

Chapter V contains a summary of the findings and some concluding remarks.

TABLE OF CONTENTS

TABLE OF CONTENTS (Continued)

LIST OF TABLES

LIST OF TABLES (Continued)

CHAPTER I

A BRIEF REVIEW OF THE LITERATURE ON THE RELATION BETWEEN GROWTH AND DISTRIBUTION

The theoretical and empirical literature investigating the relationship between income distribution and growth has usually concentrated on one aspect of this relationship, i.e. it is either analyzed how growth affects income distribution or how an existing or changing income distribution affects the growth performance of an economy.

The first strand of literature dates back of course to Simon Kuznets' presidential address to the American Economic Association in 1955 during which he presented his famous hypothesis that the pattern of inequality over time roughly follows an inverted U-shape. In his own words, "This long swing in income inequality must be viewed as part of a wider process of economic growth" (Kuznets, 1955, p. 20). The scenario of the process as well as more empirical evidence was further explored by Kravis (1960), Oshima (1962), Kuznets (1963), Adelman and Morris (1971, 1973), Paukert (1973), Chenery and Syrquin (1975) and Ahluwalia (1976 a,b). Most of these studies used cross-country regression analysis to lend support to the hypothesis.

The evidence was much less conclusive, however, as to whether there is a relationship between the level of inequality and the rate of growth of GDP (as opposed to the level of GDP). A next strand of literature therefore investigated whether the inverted U-pattern can be avoided (or at least flattened out) by appropriate policies, as well as the related issue whether the coordinates of the turning point of the U (in terms of level of GDP per capita and level of inequality) can be affected by such policies. In this context the importance of the composition of final demand has been stressed (see, for example, Ranis (1971, 1975), Adelman and Morris (1971, 1973), Land and Soligo (1971)).

Several authors have attempted to construct models which would allow the simulation of the effects of certain alternative growth strategies or policy instruments on the distribution of income. Morley and Williamson (1974) presented such a model for Brazil. A conventional input-output model, augmented by a capital requirements matrix, was used to link final demand to sectoral output. The "job competition" framework was relied upon to relate sectoral output to earnings and their distribution. A large simulation model was developed for Korea by Adelman and Robinson (1976). It consisted of a static wage and price endogenous, computable general equilibrium model, linked to a partially adaptive dynamic model. The model traced out direct and indirect influences of economic growth strategies on the distribution of income and the extent of poverty.

Finally, there is a third approach to studying the effects of growth on distribution, next to the cross-country regressions and the simulation models, viz. the historical approach, where over time changes in distribution within one country are described as the process of development takes place. Usually this approach is accompanied by attempts to break down the overall distribution into subgroups (per sector, type of income, occupation, region, etc.) in order to see how changes in the distribution within each subgroup, as well as changing differentials between groups, affected overall inequality. Formally, this procedure involves decomposing inequality measures. The list of such historical and/or decomposition studies is quite long, and a limited survey (covering about 20 such studies) is in chapter one of Grootaert (1978).

The second part of the growth-distribution relationship, namely the impact on growth of changes in the size distribution of income, is usually discussed in terms of the trade-off argument. When the composition of consumption and/or the marginal savings rate is different across income groups, then redistribution potentially affects certain constraints to growth such as savings, imports, etc. Discussions of the merits and demerits of this argument can be found in Oshima (1970), Cline (1972), Soligo and Land (1972), Looney (1975) and Grootaert (1978).

Quite a number of empirical tests of the trade-off hypothesis have been constructed with mixed findings (see, for example, Cline (1972), Chinn (1972, 1977), Figueroa (1975), Foxley (1976), Lopes (1973), Soligo (1973), Stewart (1978), Sunman (1973)). Most of these studies feed a hypothetical redistribution of income through a saving function and a set of detailed consumption functions, to obtain, respectively, the resulting change in savings available and the change in the composition of consumption. A Harrod-Domar relationship then allows to compute the effect of the change in savings on the growth rate. The change in the composition of consumption is, via an input-output table, translated into a vector of industrial output. This vector is then premultiplied with labor coefficients to yield the changes in employment resulting from the redistribution. At this point, however, the analysis is usually terminated.

Now it would be a fairly simple step to consider the income effects of the generated employment. This would indicate the induced, "second round" effects on income distribution following from an original, exogenous act of redistribution. Similarly, one could run the whole cycle of the model a few more times to see how the system works itself toward equilibrium. One could then see whether induced effects fortify or tend to cancel the original redistribution, in other words one could see whether the dynamics of the system have a built-in tendency toward equalizing or toward worsening income distribution. Apart from this quite useful result, it would also solve the technical inconsistency resulting from combining a Leontief-inverse matrix, which has equilibrium characteristics, with an otherwise short-run disequilibrium model.

A number of authors such as Cline (1972), Soligo (1973), Looney (1975) have drawn attention to the importance of including feedback effects, without, however, attempting to do so empirically. An early empirical effort that incorporates feedback effects by closing an input-output model for a disaggregated household sector is Ballentine and Soligo's study (1974) on Colombia. This study demonstrated the possibility of having indirect effects of redistribution that substantially offset the direct effects, and thus the importance of including such effects. Later empirical studies along these lines are the Paukert, Skolka and Maton (1975) model for the Philippines, and Weisskoff's study (1976) on Puerto Rico.

Essentially, including feedback effects is the basis for integrating the two approaches outlined above in studying the relation between growth and distribution. Such integration was already recommended by Kuznets (1963) when he stated, "The first question [deals with] the effects of the changes in the nature and structure of production accompanying modern economic growth on the size distribution of income..... The second question [deals with] the influence, in turn, of the size distribution of income on the process of economic growth." (p. 68). In practice, such integration needs to be done both on the supply and demand sides, but up to date no model has been designed that fully accomplishes this task.

In Chenery et al. (1975), a model for a segmented developing economy is presented that integrates the explanation of growth and distribution on the supply side, viz. via the dual role of capital. Capital is an input in the production function (and thus a determinant of economic growth) and main source of income inequality (because of the skewed distribution of the ownership of it). The model then recognizes the interdependence between growth in one income group and growth in another, because of the fact that output produced by capital owned by one group flows in part to other groups in the form of wages. It is this interdependence that is the core of an integrated theory of growth and distribution. These income group linkages "bring out the relationship between the internal dynamics of growth in each segment of the economy and the growth of income in other segments" (Chenery et al., 1975, p. 210). In Chenery's model the segments are socio-economic groups defined according to their capital ownership and income level.

Linkages between socio-economic groups also exist on the demand side. First, the spending of one household results, via production and remuneration of factors of production, in income for other households. Second, the composition of investment, government spending and foreign trade also determines the product mix in the economy and thus the flows of income to households. The model presented in this paper is intended to capture both of these channels.

CHAPTER II

AN EQUILIBRIUM MODEL OF INCOME DISTRIBUTION

II.1 Introduction

Considering the critique of existing research work as expressed in the previous pages, the model presented here should incorporate at least the following features.

First, it must recognize the existence of distinct socio-economic groups in the society, each with its own consumption and saving behavior. Second, it must incorporate as explanatory variable the level and sectoral composition of final demand, broken down in its components, viz. consumption, investment, government spending and net exports. Third, it must be able to generate both the level and distribution of household income.1/ Fourth, it must incorporate feedback effects from the level and distribution of income (itself generated by a given amount and pattern of spending), back to spending, then back to income, etc. In other words, the existence of a multiplier mechanism of spending on household income must be incorporated in the model.

Socio-economic groups can be defined according to several criteria, such as income level, family size and composition, location (region, urban/rural), occupation or industrial affiliation of the household head, etc. The criteria to be selected for a particular model application are in principle determined by the purposes of the analysis, although in practice they will also depend on the nature and content of the available data source.

1/ The purpose of our analysis is to relate changes in the structure of final demand to changes in the degree of (in)equality of income distribution. We are not concerned here with equity which is a concept requiring value judgments as to the welfare implications of (changes in) income distribution.

The inclusion of feedback effects can be accomplished in two ways.

The first way is by making the model dynamic. In reality, a lag structure does indeed exist between spending and production, production and income, income and spending.

The lag between income and spending has substantial theoretical basis in, for example, the permanent income and the habit hypotheses.1/ A lag also exists between the production of industrial output and the resulting income flow to households. The structure of this lag is bound to be quite complex in reality. For profits it will depend mostly on the time pattern of dividend payments. For wages the lag may be quite short or even negative since, in terms of cash flows, wages are often received by households even before the proceeds from sales of finished goods are obtained by the firm.

A third place where in reality time elapses is in the production process itself. The relation between demand and output will in our model be represented by an input-output table. An input-output model, however, is a static equilibrium model.2/ It incorporates intersectoral feedbacks, but without considering the time necessary for them.

This leads to the second way feedback effects can be included in a model, viz. by making it an equilibrium model. An equilibrium model differs in at least two ways from a dynamic model: it does not show the adjustment path towards equilibrium, and it gives no indication of the time needed for the adjustment. Otherwise, both types of models do incorporate "second round" effects in their solutions.

1/ For a discussion, see for example, Wonnacott, 1974, pp. 348-352.

2/ At least the simple version, when no capacity expansion matrix is included.

Considering the fact that we are primarily interested in the importance of feedback effects rather than in the actual time elapse involved, we opted for an equilibrium model.1/ The time dimension is lost, but, on the other hand, there is a gain in internal consistency, as all model components have the same equilibrium nature as the input-output core of the model.2/

We now proceed with the formal presentation of the model.

II.2 An Equilibrium Model of Income Distribution

Formally, the model is derived from the general equilibrium condition that aggregate demand equals aggregate supply for each sector.

Aggregate supply is the sum of domestic output and imports. Aggretate demand is broken down in three components: intermediate demand, consumption (both of these are endogenous to the model) and exogenous final demand. We can thus write:

1/ In addition, the data that are available for the empirical application in this paper are mostly cross-sectional and it would therefore require a number of fairly heroic assumptions to estimate the lag structures for a dynamic model.

2/ This is more important than merely to achieve mathematical homogeneity. Note, for example, the oddity in interpreting the heuristic dynamic income distribution model presented by Cline (1972): final demand is a function of level and distribution of income in the previous time period. Total output though is a function of final demand in the same period, in spite of the fact that the link between the two is the Leontief-inverse matrix. While formally this specification is correct (the Leontief-inverse is a static model component), economically, in the framework of an otherwise dynamic model, it seems inconsistent to claim a time-lag between income and spending, while claiming no lag between spending and production. Little confidence could be placed in the time-path of income distribution traced out by such a model. The distinction static vs. dynamic in mathematical sense does not always coincide with the same distinction when made in the real world. This point cannot be overlooked in the interpretation of model results.

$$X + M = AX + C + FD \qquad (1)$$

where X is a $j \times 1$ vector of domestic output (j is the number of input-output sectors)

M is a $j \times 1$ vector of imports

A is a $j \times j$ matrix of technical coefficients

C is a $j \times 1$ vector of household consumption

FD is a $j \times 1$ vector of exogenous final demand 1/

It is one of the explicit model requirements to distinguish different socio-economic groups, each of which has its own spending pattern. This is represented by

$$C = cE \qquad (2)$$

where c is a $j \times i$ matrix of consumption coefficients

(j is the number of spending categories) 2/

(i is the number of socio-economic groups)

E is a $i \times 1$ vector of total household spending.

Each element c_{ij} of matrix c represents the propensity to consume spending category j by the ith socio-economic group. 3/ The elements of vector E are the average total expenditures of each socio-economic group.

1/ Actually FD represents the sum of n column vectors, where n is the number of final demand components distinguished.

2/ The reader will note that j represents both input-output sectors (see equation (1)) and spending categories. The nature of our model makes it necessary to group expenditure items into categories corresponding with input-output sectors. The adjustments and assumptions involved in this process are discussed in section III.2.1 of this paper.

3/ The estimation procedure of the propensities to consume will be described in section III.2.1 of this paper; a discussion as to whether average or marginal propensities are most appropriate can be found in section II.4.

Combining (1) and (2) yields, after re-writing,

$$X - AX - cE = FD - M \tag{3}$$

Next, we consider the income generation process, which we write as

$$Y = VX + \bar{Y} \tag{4}$$

where V is a $i \times j$ matrix where each element v_{ij} represents the share of value added generated in sector j that is received by socio-economic group i

$\bar{Y}$ is a $i \times 1$ vector of exogenous income (for example transfer payments).

Equation (4) can be re-written as

$$-VX + Y = \bar{Y} \tag{5}$$

We close the model by noting that

$$E = eY \tag{6}$$

where e is a $i \times i$ diagonal matrix where the non-zero elements are the ratios of household expenditures to income for each socio-economic group

We can combine (3), (5) and (6) to yield

$$\left[\begin{array}{c|c} I - A & -ce \\ \hline -V & I \end{array}\right] \left[\begin{array}{c} X \\ \hline Y \end{array}\right] = \left[\begin{array}{c} FD - M \\ \hline \bar{Y} \end{array}\right] \tag{7}$$

where I is the identity matrix.

Writing $c' = ce$, the solution of (7) is

$$\left[\begin{array}{c} X \\ \hline Y \end{array}\right] = \left[\begin{array}{c|c} I - A & -c' \\ \hline -V & I \end{array}\right]^{-1} \left[\begin{array}{c} FD - M \\ \hline \bar{Y} \end{array}\right] \tag{8}$$

Using the formula for inverting a partitioned matrix (Goldberger, 1964, p. 27-28), we obtain:

$$\begin{bmatrix} X \\ \hline Y \end{bmatrix} = \left[\begin{array}{c|c} B(I - c'VB)^{-1} & Bc'(I - VBc')^{-1} \\ \hline (I - VBc')^{-1} VB & (I - VBc')^{-1} \end{array}\right] \begin{bmatrix} FD - M \\ \hline \bar{Y} \end{bmatrix} \qquad (9)$$

where $B = (I - A)^{-1}$.

Expression (9) describes the generation of the distribution of income over socio-economic groups in a manner fully consistent with the structure of production. The consistency is achieved because the model represents a closed-looped system (the closing is done by making consumption endogenous) in which "production leads to factor earnings, which leads to income determination, which leads to expenditure patterns, which leads to production requirements needed both to supply the goods and to create the income flows" (Clark, 1975, p. 142).

Regarding distribution, expression (9) is flexible in that Y can contain distribution over socio-economic groups defined along any one criterion or set of criteria. The only restriction is that the rows of V and the columns of c' must be defined along these same criteria (in order to effectuate model closure).

We shall now proceed with a discussion of the solution matrix in equation (9).

II.3 The Solution Matrix

II.3.1. The Linkages Matrix

The dominant feature of solution equation (9) is the submatrix $(I - VBc')^{-1}$, since it appears in each part of the solution matrix. 1/ Miyazawa (1976) calls this submatrix the "interrelational income multiplier" (p. 6). It has dimensions $i \times i$, and if we call its elements $z_{k,\ell}$, where $k = 1,\ldots,i$, and $\ell = 1,\ldots,i$, then each element represents the income received by socio-economic group k as a result of the spending by group ℓ out of one unit of income. The elements of submatrix $(I - VBc')^{-1}$ are multipliers, representing a three-step propagation process.

The first step, represented by c', is the consumption spending out of earned or received income. The second step, captured by the Leontief-inverse matrix, "translates" the spending into industrial output. The third step, V, tells how much income is generated for each socio-economic group by the production of this output. This newly generated income results in new spending, and the process repeats itself until an equilibrium position is reached (the latter is guaranteed by the inversion procedure).2/

Submatrix $(I - VBc')^{-1}$ thus portrays how each socio-economic group is linked to every other group via the processes of consumption, production, and income generation. We will refer to this matrix as "matrix of inter-socio-economic group linkages" or, briefly, "linkages matrix".

1/ It can be shown that $B(I - c'VB)^{-1} = B(I + c' (I - VBc')^{-1}VB)$.

2/ In addition to the conditions applying to the Leontief-inverse, the convergence of the propagation process is guaranteed if and only if for each socio-economic group the consumption coefficients add up to less than one (see Miyazawa, 1976, p.15-21).

It is this kind of linkage that Ahluwalia and Chenery (1975) refer to in the context of integration of growth and distribution models. The reader will recall from the discussion in Chapter I, that in their own model the link was provided by the stock of capital in its dual role of determinant of growth (because it is input in the production function) and of determinant of distribution (because of its ownership concentration in the rich group).

In our model, the link is provided by endogenous consumption demand, broken down by socio-economic groups. This incorporates the feedback effects in the model which allow to relate the spending of one socio-economic group to the income of another group.1/ In addition, since our model explicitly incorporates the sectoral composition of final demand, the linkages between socio-economic groups are based not only on the amount of consumption but also on the pattern of consumption.

The row and column totals of the linkages matrix provide important information. The column totals are total income multipliers, i.e. they show the total household income generated by a unit of income originally earned by each socio-economic group. The column totals therefore indicate how total household income will change if income is redistributed across socio-economic groups.

1/ It should be noted that, if one does not distinguish between socio-economic groups, the linkages matrix reduces to $\frac{1}{1-e}$, where e is the aggregate marginal propensity to consume, which is, of course, the standard Keynesian multiplier (see Miyazawa, 1976, p.10-12).

The row totals of the linkages matrix show the total income received by each socio-economic group, if each group were to earn (and spend out of) one unit of income. A particularly useful way of interpreting the row totals arises if we define socio-economic groups by the income criterion, i.e. if we equate socio-economic groups with income groups.[1/] Let us say, for the sake of easy explanation, that we distinguish five income groups each with equal population shares, i.e. quintiles. Suppose then that marginal incomes are distributed perfectly equally (i.e. any additional incomes are distributed 20% to each quintile), then the row totals represent the resulting equilibrium distribution of that marginal income. It is, in other words, the minimum level of inequality feasible, given the three structural elements of the economy captured by our model, viz. the consumption and saving behavior of households, the input-output technology, and the generation and distribution of households' share of value added.[2/]

The inequality of the actually generated income distribution will be above this minimum due to the existing (exogenous) inequality of income distribution and because the composition of exogenous final demand will not usually be so that inequality of distribution is minimized.

1/ This is in fact what we shall do in the empirical application in Chapter III, for reasons of data availability explained there.

2/ It is assumed that there are no continuous redistribution programs which give, at the margin, the lower income quintiles each more than 20% of income. If there is such a program, inequality of income distribution could fall below the level predicted by the linkages matrix.

The linkages matrix provides the basis for breaking down the actually observed level of inequality into a component resulting from the three above mentioned structural elements - we shall refer to this component as "endogenous inequality" - and a component resulting from exogenous elements. This distinction has important policy implications. A transfer program, for example, which does not change the structure of taxes, will affect the exogenous level of inequality but not the endogenous one. If the latter is on a level fairly close to the former, then the equilibrium effect of a transfer program wil be minimal with respect to distribution, and policy action should be focused on structural changes. If, on the other hand, the level of endogenous inequality is far below the exogenous one, there is more scope for exogenous marginal redistribution.

The linkages matrix can also be used to explore the steady state income distribution that is consistent with the three structural components captured by our model. Formally, this can be represented by the eigenvalue problem

$$(I - VBc')^{-1} \; y = \lambda y \tag{10}$$

where y is an eigenvector of the linkages matrix and λ its associated eigenvalue.

For y to represent an income distribution vector, specifically, a vector of population quintile income shares, the sum of its elements must equal one, i.e. for $y = [y_i]$, $\sum_{i=1}^{5} y_i = 1$, and each element of y must be non-negative, i.e. $y_i \geq o$ for each i. If we call the vector that meets these conditions s, then (10) can be re-written as

$$(I - VBc')^{-1} \; s = \gamma s \tag{11}$$

where γ is the new associated eigenvalue. Note that (11) eliminates the possibility of the trivial solution of a zero eigenvector. This implies that $(I - VBc')^{-1}$ must be non-singular if (11) is to have a solution.

The linkages matrix is obviously square and will usually be non-negative. A negative element would mean that an increase in the income of a socio-economic group would reduce the income of another group. Since the elements of V and B cannot be negative if they are to be economically meaningful, this could only occur if elements of c' are negative, i.e. if some groups have negative marginal propensities to consume for certain items. Although this is possible it is not likely to result in negative elements in $(I - VBc')^{-1}$ due to interactions with the other components. Whether the linkages matrix is indecomposable will depend on the level of aggregation. At quintile level it is bound to be. It will be decomposable if there is at least one group in society whose spending feeds back in no way, neither via production nor value added generation, into the income of another group. This could occur if a large number of highly disaggregated groups are considered, for example, broken down regionally in a situation where there is little inter-regional trade. The normal case, however, will not present a decomposable matrix.

If, then, the linkages matrix $(I - VBc')^{-1}$ is square, non-negative and indecomposable, the Frobenius-Perron theorem applies which ensures that the matrix has a dominant root (a unique largest eigenvalue) with which is associated one and only one strictly positive eigenvector. Therefore, for most economic applications and at normal levels of aggregation, equation (11) will have a unique solution vector that is economically meaningful and represents a steady state income distribution. The steady state interpretation can perhaps better be seen from the fact that s can also be obtained non-algebraically, viz. by an iteration process as follows. Let s_1 be the initial distribution vector, then

$$d_2 = Ls_1$$

where $L = (I - VBc')^{-1}$. The elements of the resulting vector d_2 will not sum to one. If we therefore normalize d_2, the resulting distribution vector s_2 is the basis for the second iteration:

$$d_3 = Ls_2 \quad \text{where} \quad s_2 = \left[\frac{d_{2,i}}{\sum_i d_{2,i}}\right]$$

This process can be continued until $d_n = Ls_{n-1}$ where $s_{n-1} = \left[\frac{d_{n-1,i}}{\sum_i d_{n-1,i}}\right]$.

For $n \rightarrow \infty$, $s_n = s_{n-1}$, where $s_n = \left[\frac{d_{n,i}}{\sum_i d_{n,i}}\right]$. 1/

This result is independent of s_1, and s_n (which is equal to s in equation (11) represents therefore the distribution that would maintain itself indefinitely if no structural changes would occur. 2/

It is worth noting that s can also be obtained from an eigenvalue equation involving VBc' rather than $(I - VBc')^{-1}$. This can be seen by pre-multiplying (11) by (I - VBc') which yields

$$s = (I - VBc')\gamma s$$

$$s = \gamma s - VBc'\gamma s$$

$$(1-\gamma)s = -VBc'\gamma s$$

$$\left(\frac{\gamma-1}{\gamma}\right)s = VBc's \qquad (12)$$

1/ The normalization of the successive vectors d to s is not required in practice; one normalization after the nth iteration will produce the same result.

2/ Although the steady state distribution will be arrived at from any initial distribution, the time needed to reach it may be affected by the starting point.

In other words, VBc' has a different dominant root but the same associated eigenvector. If VBc' meets the conditions specified for the Frobenius-Perron theorem to hold, the vector will be unique and strictly positive.

It should also be clear that s can be obtained equally from the full structural matrix of equation (7), and, alternatively from its inverse in equation (9). For the structural matrix, the eigenvalue problem is:

$$\left[\begin{array}{c|c} I - A & -c' \\ \hline -V & I \end{array}\right]\left[\begin{array}{c} x \\ \hline y \end{array}\right] = \left[\begin{array}{c} \lambda_1 \\ \hline \lambda_2 \end{array}\right]\left[\begin{array}{c} x \\ \hline y \end{array}\right] \tag{13}$$

Here, constraints are to be placed separately on y and x so that all their elements are positive and sum to one.

Consideration of the steady state level of inequality and the gaps between it and actual and endogenous inequality has important policy implications. Lowering steady state levels requires policy action geared towards structural changes such as redistribution of assets, changes in technology, etc., while income redistribution, altering the composition of investment or government spending, etc. may be used to reduce the gap between actual and steady state levels. If funds are redistributed perfectly equally at the margin, then the endogenous distribution will prevail initially, but the long-run effect will be to worsen that distribution again, and the more so the wider the gap between endogenous and steady state inequality.

We would argue that an international comparison of inequality levels could meaningfully be done by comparing linkages matrices and their implied levels of endogenous and steady state inequality. This would yield more insights than merely comparing total observed inequality. Countries can indeed have similar observed levels but widely varying steady state and/or endogenous levels, and vice versa. Obviously this has important implications for the correct understanding of the processes underlying cross-country distributional patterns. It would be interesting to explore, for example, whether the Kuznets curve is also observed for endogenous and/or steady state inequality, and whether the size of the gaps between the different levels of inequality is related to the level of development.

II.3.2 The Other Submatrices in the Solution

We shall now take a look at the three other submatrices of solution equation (9).

Submatrix $(I - VBc')^{-1}VB$ is the linkages matrix post-multiplied with VB. It translates final demand into income for each socio-economic or income group. This is done in two steps: first, the total effect of final demand on production is considered (this is B[FD - M]) and its implication on household income (this is VB[FD - M]).

Up till this point no feedback effects from the newly generated incomes on consumption are included. This is the point of the analysis where most studies on income distribution that include demand aspects stop (These studies were surveyed in Chapter I). As we argued earlier, limiting the analysis to the matrix VB[FD - M] disregards the second and higher round effects of income on consumption and presents an internal inconsistency in the model as it does incorporate the intersectoral feedback effects implicit in the Leontief-inverse, while omitting those of income on consumption. This inconsistency is in our opinion a serious one as in reality the time needed for the interindustry feedbacks to work themselves out is no doubt much more than the time required for the income to consumption feedback effects.

Our model therefore adds a second step to the translation of final demand into income by multiplying VB[FD - M] with the linkages matrix. The resultant matrix $(I - VBc')^{-1}$ VB contains multipliers expressing the direct and indirect effects of a one unit increase in final demand (per input-output sector) on each socio-economic or income group.

It holds the key to explaining how shifts in the sectoral composition of final demand affect income distribution. Among many applications possible, this matrix would be extremely useful for project analysis in developing countries as it would give directly the long term implications for household income and its distribution of any project.[1/]

This brings us to the third submatrix of equation (9), viz. $B(I - c'VB)^{-1}$ which gives the production requirements per unit of spending. It is the equivalent in our model of the simple Leontief-inverse in a model without distributional implications. The relationship with the latter is clear as it is the Leontief-inverse post-multiplied with $(I - c'VB)^{-1}$. In other words, the model first explores the production-income-consumption requirements of a unit of spending and when the direct and indirect effects on consumption are found, they are then, together with the original unit of spending, traced through the Leontief-inverse to obtain the total effect on production. Again, it would seem that the use of this matrix in project analysis would be superior to the use of the simple Leontief-inverse in order to obtain the sectoral output requirements of any given project.

The last submatrix of expression (9) is $Bc'(I-VBc')^{-1}$, giving the total output requirements per sector of the spending resulting from a one unit of income of each socio-economic or income group. Rather than directly considering the production effect of the spending of each group (which would be Bc') the submatrix first considers the total income effects (i.e., $(I - VBc')^{-1}$) and only then computes consumption associated with that total income and its production requirements.

1/ The selection of investment projects as a tool to influence income distribution is recognized by lending agencies such as the World Bank, especially when the government lacks the political power to directly alter the existing income distribution, as is the case in many LDC's. See Lal (1972), Turnham and Hawkins (1973).

The following table summarizes the relationships captured by each submatrix of solution equation (9).

effect of / on	spending (per sector)	income (per socio-economic group)
output (per sector)	$B(I - c'VB)^{-1}$	$Bc' \ (I - VBc')^{-1}$
income (per socio-economic group)	$(I - VBc')^{-1} \ VB$	$(I - VBc')^{-1}$

II.4 Assumptions of the Model

Since the model is an expanded input-output model, it is subject to all the assumptions and limitations inherent in the input-output technique.

The fixed coefficients imply infinite elasticity of supply of all factors of production. During the adjustment process towards equilibrium, no factor substitution is assumed to take place.

Since the model recognizes the existence of different socio-economic groups, it implicitly recognizes the heterogeneity of the labor force. The fixed coefficients then also imply that no substitution takes place between different skill groups as a result of changes in relative wages.

This approach is roughly consistent with the "job competition" model in which job characteristics and vacancies (rather than formal education) are the basic determinants of market earnings. It implies the existence of heterogeneous labor skills which are assumed for the most part to be acquired on the job. Workers with background characteristics which yield minimum expected training costs will be selected first.

The job competition framework is relevant where on-the-job training is important, where strong seniority provisions exist, and where competitive forces in the labor market are weak so that relative wages are not very important in allocating labor (see Morley and Williamson, 1974). With respect to our empirical application (see next chapter), we believe that these three conditions are fairly well met in the Japanese case, at least in the fifties and early sixties. Later in the sixties the labor market in Japan became more tight and the role of competitive forces in bidding up wages and affecting labor allocation became more important (see section III.1).

The Matrix V describes how value added in each sector is allocated to the different socio-economic groups. As such it represents the so-called "primary income distribution," which is the distribution "directly established by flows of income shares from the production system" (Kuznets, 1975, p.790). It results directly from the receipt by households of wages and income from assets and is closely related to and affected by the production structure and by growth-related changes in the production structure.

Our model distinguishes the primary income distribution from the secondary income distribution which is the one that obtains after government intervention via taxes and transfers, and after capital gains and various shifts of assets between households have taken place. It is the secondary income distribution that affects the production structure because it determines the spending patterns of households. Therefore, when describing the feedback mechanism from production to income to consumption, the model breaks the income link down into two steps, viz. production to primary income distribution, and secondary income distribution to consumption. Formally, this is achieved by equation (6), which takes taxes out of incomes received by the households in each socio-economic group.[1/]

The treatment of imports in the model also deserves some comment. Imports are assumed to be exogenous and treated "as if" they were all of the competitive type. This does not mean that all imports need to be produced domestically as well, but merely refers to the fact that imports are a column vector in the final demand section of the model.[2/]

Treating imports as exogenous implicitly assumes that the ratio of imports to gross output is constant over time for each sector. It is considered a major weakness of the input-output methodology that it keeps all coefficients constant over time.[3/] In our research design this is not much of an issue since the model is not used for explicit over-time predictions. However, as was discussed earlier in this chapter, we do give an implicit time-dimension to an equilibrium model, viz. the time it would take in reality for all the second-and-higher-round effects to occur until an equilibrium is reached. So, what is required is that the technical coefficients, the consumption coefficients, the import ratios and the value added coefficients are constant during that time.

Our analysis in later chapters will be based on a comparison of model solutions for two selected years. For each of these years we have a complete

---

1/ Equation (6) also takes saving out of incomes received because the consumption coefficients in matrix c have total household expenditures, rather than disposable income, as argument.

2/ The Japanese input-output tables used in our empirical application treat imports this way (see Link Input-Output Tables, Explanatory Report, 1975, p.119-122) and we have simply taken over the methodology. Note that if imports are treated as noncompetitive, they form one or more additional rows to the matrix of technical coefficients.

3/ See for example Arrow and Hoffenberg (1959); Chenery and Clark (1959), especially Chapter 6; Leontief (1953) especially Chapters 2 and 3.

set of coefficients, hence for explicit over-time comparison, we do not make the assumption of constant coefficients.

A related issue is whether it is appropriate, as input-output methodology does, to use average coefficients. It is our opinion that it is not. As discussed earlier, the model solution is the result of several rounds of adjustment between income, spending and production. Consequently, marginal coefficients are appropriate, as indeed one must describe how additional income is spent, how this additional spending affects production, and how additional production generates household income. We shall therefore estimate marginal consumption coefficients for each socio-economic group (the details of this procedure are in section III.2.1). For the technical coefficients and the value added coefficients though, the data available for the empirical application necessitated the assumption that the marginal coefficients equal the average ones. Since in Japan technology changes were rapid, and changes in the labor market substantial in the late sixties, the assumption is bound to be violated. Still, the reader is reminded that the assumption again only applies to the time needed implicitly for the model to reach equilibrium.

CHAPTER III

AN APPLICATION TO JAPAN

In this chapter we are applying our model to Japanese data spanning the period 1959-1969. During this decade economic growth in Japan was steady and uninterrupted, providing an opportunity to clearly observe the effects of growth related changes in final demand on distribution.

The first section of the chapter reviews the trends in income distribution during those years. In section II the data base and the estimation procedures of the model coefficients are discussed. Section III contains the complete model solutions for each of the two years of analysis, and section IV summarizes the findings.

III.1 Studies on the Japanese Income Distribution

In the immediate postwar years, the income of the highest income class in Japan decreased, while the income of every other class increased. Income distribution in the salary and wage earners group improved in prosperity years (1948, 1951) but worsened in recession years. From the mid-fifties income inequality increased until 1961-62. Subsequently, the trend was reversed, and inequality fell. From the late sixties on though an upward tendency of the Gini-coefficient has been noted again (Takahashi (1959), Ishizaki (1967), Mizoguchi (1975a, 1975b), Mizoguchi, Takayama and Terasaki (1977), Wada (1975)). Table III.1 shows the time pattern of the Japanese income distribution between 1954 and 1971.

Table III.1

Time-Pattern of the Japanese Income Distribution

Year	(1)	(2)	(3)	(4)	(5)	(6)
1954	.224					
1955	.225					
1956	.215				.276	.313
1957	.229					
1958	.227					
1959	.228				.319	.357
1960	.234					
1961	.236					
1962	.226	.2744	.3629	.3759	.340	.382
1963	.225	.2544	.3461	.3607		
1964	.215	.2440	.3402	.3528		
1965	.201	.2392	.3276	.3441	.337	.377
1966	.213	.2373				
1967	.212	.2403	.3307	.3523		
1968	.200	.2191	.3257	.3488	.324	.380
1969	.189	.2017	.3319	.3539		
1970	.188	.2197	.3284	.3553		
1971	.189	.2094	.3301	.3521	.347	.407

(1) Gini-coefficient for pre-tax income of employee households based on the Family Income and Expenditure Survey (FIES) (figures are approximate, as read off a diagram). Source: Mizoguchi (1975a), Diagram I, p. 480.

(2) Gini-coefficient for pre-tax income of all multi-member households based on the Family Income and Expenditure Survey (FIES), the Cost of Living Survey for Farm Households (CLSF), and the Family Saving Survey (FSS).

(3) Gini-coefficient for pre-tax income of multi-member households based on the Survey of People's Living Conditions (SPLC).

(4) Gini-coefficient for pre-tax income of total households based on the SPLC.

(2) - (4) Source: Mizoguchi, Takayama and Terasaki (1977), Table 2, p. 12.

(5) Gini-coefficient for pre-tax income of multi-member households based on the Employment Status Survey (ESS).

(6) Gini-coefficient for pre-tax income of all households based on the ESS.

(5) - (6) Source: Wada (1975), Table 4.5-2, p. 77.

Explanations of these trends have focused on institutional and demographic changes and on changes in the labor market. The equalization of income in the late forties has been explained by the dissolution of the Zaibatsu, the agrarian reform, and the institution of a property tax. In the fifties, the share of property income in total income rose and property income became more concentrated again in the higher income classes. At the same time, wage differentials between large and small enterprises increased rapidly. The relative unequal distribution of bonus payments to employee households also played an important role in explaining the regressiveness in the trend of income distribution in the fifties. After 1955 there was a rapid increase in the number of one person households. Since these households were mostly young persons ready to enter the labor force or retired people with a fairly low income, this was a further cause of widening income inequality (Ishizaki (1967), Wada (1975), Boltho (1975), Mizoguchi, Takayama and Terasaki (1977)).

In the early sixties, the labor market shifted from surplus to shortage, and wage differentials diminished. This is generally seen as the major reason for the decline in inequality in the sixties (Ishizaki (1967), Wada (1975), Minami (1973), Ono and Watanabe (1976)).

Among multi-member households, average household size declined, while at the same time the average labor force participation rate increased. Particularly in low income households the number of secondary earners increased. The result was that secondary earners offset the inequality among principal earners and thus contributed to a more equal distribution of income among households. Furthermore, counter to what happened in the fifties, the weight of property income which is relatively unequally distributed increased little in the sixties, so that this aspect of the composition of household income had little influence on the overall distribution (Wada (1975), Mouer (1973, 1974)).

Agricultural incomes were increasingly supplemented by income from non-agricultural activities. In combination with the rice price support programs of the government, this tended to equalize agricultural income distribution until 1967 and narrow the rural-urban income gap. After 1967, inequality in agriculture rose, as the taking of non-agricultural side-jobs spread to higher income farm households and as the government changed its policy to promoting non-rice farming, which benefited larger scale farms (Mizoguchi (1975b), Ono and Watanabe (1976)).

As for the distribution of non-agricultural incomes, analysis of variance revealed that inequality within occupation, age and firm-size groups and its trends over time were major explanatory factors of the trends in the overall distribution (Mizoguchi (1975b), Cole and Tominaga (1976)).

Finally, rural-urban migration was found to be a factor promoting equality by reducing income gaps between city sizes, and ultimatley, as labor scarcity became more widespread, also between cities and rural areas (Mizoguchi, Takayama and Terasaki (1977)).

Boltho (1975) points at several socio-economic facts typical to Japan which can help explain its low level of inequality. Specifically, he mentions the virtual absence of three groups of people who form the bulk of the poor in Western countries, viz. racial minorities, immigrants, and the old. The latter live, in accordance with the traditional (albeit rapidly disappearing) family structure often with their children, rather than in independent households as is the case in the West. [1] These factors are reflected in the fact that the income

1/ Still, the low income levels of many households of old people, of fatherless families, and of households without employed workers form a potential target for Japanese redistributive efforts (Economic Planning Agency, Economic Survey of Japan 1971-72, pp. 133-134).

share of the top 5 or 10% of Japanese households is not much different from that in other "egalitarian" societies like Australia or Norway. The difference is found in the bottom 20% of Japanese households who receive close to 8% of total income as compared to 4-6% elsewhere.

Boltho further observes the peculiar phenomenon in Japan that the share of participation income (= employee compensation plus income from unincorporated enterprises) in national income has declined over time, while the share of income from assets has risen over time. Possible explanations include structural shifts (given different factor shares in different sectors), the gradual decline of the unincorporated sector where capital share is below average, and the increasing share of investment in output (which, in a Kaldorian model, would indeed predict a rising share of profit in national income).

Boltho concludes that "Japan's particular income distribution pattern seems to have combined the 'best of both worlds'." (p. 183). The equal personal income distribution was favorable to welfare. The bias in functional distribution against participation income was favorable to growth, since, given the strong reinvestment tendency of Japanese corporations, it provided for a large and growing share of income for capital accumulation. "Japan seems to have avoided, at least in part, the dilemma of reconciling equity and efficiency which faces so many of today's developing countries" (p. 183). The key elements in this are high saving propensities at all income levels as well as a high saving ratio out of corporate profits. Under those conditions equality in personal income distribution is no hindrance to provide the amount of saving needed for investment needed for growth.

The studies surveyed above emphasized demographic and institutional factors and changes in the labor market. None of them investigated the role of growth related changes in the structure and composition of final demand. Such investigation is the purpose of the model application in this chapter, and several new insights into the process of a changing income distribution will be gained from this. Also, an evaluation of the importance of changes in final demand relative to labor market related changes will be possible.

III. 2 Data Base and Estimation of the Coefficients of the Model

III.2.1 The Consumption Coefficients

We need to estimate consumption coefficients (Engel coefficients) for each socio-economic group distinguished in the model. It is therefore necessary at this point to decide on the criteria to define these groups for the purposes of this application of the model. Ideally, with access to the individual questionnaires of a household survey, analysis of variance techniques could be used to find out the relevant classification criteria. For the purposes of this paper, we had to take the published results of the Japanese household surveys as ultimate data source.

While this still provides for a multitude of classifications, a further limitation is imposed by the requirement that the model must incorporate feedback effects from the spending to the income of each socio-economic group. This requires that data are available about the flows of income generated in each (input-output) sector of the economy to each socio-economic group. Such data were only available for income groups, and hence we were restricted for this application to using only the income dimension of socio-economic groups. We are thus in fact starting from the assumption that income groups are relevant socio-economic groups with distinctly different consumption and saving behavior. Our econometric analysis will, of course, have to confirm this assumption.

The data source for the estimation of the consumption coefficients is the National Survey of Family Income and Expenditures (NFIE) published every five years since 1959 by the Bureau of Statistics, Office of the Prime Minister, Japan. This publication contains tabulations of average monthly receipts and disbursements per household by income group cross-classified by occupation of the household head, and so permits the estimation of marginal propensities to consume (for each spending category) within each income group, since multiple observations are available within each group. The major advantage of the NFIE over other publications is its large sample size, which makes the cross-classification possible while at the same time keeping sufficient observations in each category.

The major drawback of the survey is that the survey period extends only from September to November. We therefore compared the estimates of the Japanese income distribution made by Wada (1975) on the basis of several surveys with annual observation periods with the income distribution derived from the NFIE data, and we found that the correspondence is quite close (see Table III.2). The NFIE data have some bias toward equality, suggesting that seasonal payments (essentially bonus payments) are distributed a bit more unequally than regular monthly incomes.

We also compared, for selected income groups, the spending patterns reported in NFIE with those reported in the Annual Report of the Family Income and Expenditure Survey (FIES). The latter distinguishes between the average spending computed over the period January-November and the spending pattern in December, when the bonus is received by most households. 1/ In general the NFIE spending

1/ The data permit us to make this comparison only for urban workers' households.

Table III.2 Income Distribution, 1959, 1969

Quintile	Wada 59	NFIE 59		Wada 68	NFIE 69	
	%	$\bar{Y}$[a]	%	%	$\bar{Y}$[a]	%
I	7.9	13,204	8.1	7.3	41,959	8.5
II	12.5	21,270	13.0	12.2	63,942	12.9
III	16.6	27,825	17.0	16.2	80,276	16.2
IV	21.9	36,349	22.1	22.1	101,903	20.6
V	41.1	65,196	39.8	42.2	206,806	41.8
G[b]	.3032	.2897		.3188	.2972	
K[b]	.2875	.2747		.3038	.2798	

[a] Average monthly income per household.

[b] G = Gini-coefficient
K = Kuznets measure of inequality

Source: Wada (1975) p, 248, 251.
59 NFIE, Vol. I, Table 1, p. 31-32.
69 NFIE, Vol. I, Table 2-1, p. 56-57.

percentages differ only slightly from the January-November FIES estimates and fall between the latter and the December pattern of FIES. In the majority of cases they are close to the percentage that obtains from a weighted average of the January-November and the December pattern. We may conclude that the 3-month survey period of NFIE does not result in any major income group specific biases, and that, therefore, the NFIE qualifies as a data source for our model.

The structure of the model requires that the categories for which consumption coefficients are estimated coincide with the input-output categories. This required the subdividing and re-arranging of several NFIE categories. Also, a number of conceptual adjustments were necessary, mainly to insure appropriate treatment of credit purchases, trade margins and insurance premiums. Details are in Appendix One.

Once the NFIE categories were re-arranged and compatible with the input-output categories, the regression estimation of the marginal propensities to consume could be started. The consumption package purchased by an income group differs from that purchased by another group in two basic ways: the pattern of consumption differs across income groups, and the total level of expenditure differs. In order to estimate the consumption coefficients for our model, we need a set of sectoral consumption functions (Engel Functions) that capture both of these aspects. In addition, the functions must fulfill the additivity condition, i.e. all Engel coefficients need to sum up to one minus the propensities to save and to pay taxes within each income group.

There are several functional forms that will meet these conditions. The simplest is

$$C_{ij} = c_{ij}\, C_i \qquad \text{where } c_{ij} = \frac{C_{ij}}{C_i}\,. \tag{14}$$

This amounts, of course, to simply computing average propensities to consume out of total expenditures. This can be defended when there is some reason to believe that the Engel functions are straight lines through the origin, in which case average and marginal propensities to consume are equal. Theoretically, only the permanent income hypothesis provides a framework where the long-run aggregate consumption function meets that condition. Even in this framework though there is a no *a priori* reason to expect every Engel function to be a straight line without intercept. 1/

Since our model is an equilibrium model which includes successive rounds of adjustment, there are, as discussed earlier, theoretical reasons to prefer the use of marginal propensities to consume. Since in addition, there is no empirical evidence in the Japanese case to argue for the equality of marginal and average propensities, we reject functional form (14).

The simplest functional form of Engel functions that yield marginal propensities to consume (MPC) and allow to estimate inter-income-group differences among them, is the following:

$$C_j = a_j + b_j\, Y + \sum_i c_{ij}\, D_i\, Y \quad . \tag{15}$$

Expression (15) represents a family of j linear Engel functions (one for every expenditure category j) with dummy variables D_i for i income groups. 2/ From this expression a matrix of consumption coefficients can be derived with $MPC_{ij} = b_j + c_{ij}$ as elements.

1/ The study on the relationship between distribution and growth by Paukert, Skolka and Maton (1975) uses, however, average propensities to consume.

2/ For example, $D_1 = 1$ when $i = 1$, $D_1 = o$ otherwise. In actual estimation, one dummy variable would be dropped, of course, to avoid perfect collinearity with the intercept.

An alternative functional form which also estimates MPC's and their differences across income groups is the quadratic Engel function:

$$C_j = a_j + b_j\ Y + c_j\ Y^2 \tag{16}$$

where $MPC_j = b_j + 2\ c_j\ Y$. Here MPC itself is a function of income level. This poses a problem if we were to incorporate this functional form in the model, as a solution by simple matrix inversion is no longer possible. Instead an iterative search procedure is needed which establishes the equilibrium levels of MPC and Y simultaneously. One way to avoid this, however, is to compute MPC at the average income level $\bar{Y}$ of the relevant income group i, so that $MPC_{ij} = b_j + 2\ c_j\ \bar{Y}_i$.

As b_j measures slope, and c_j measures curvature of the Engel function, it is clear that differences in b_j and c_j between income groups are additional possible reasons (i.e. other than differences in $\bar{Y}$) why MPC's can differ across income groups. Consequently, we will also estimate the following forms:

$$C_j = a_j + b_j\ Y + c_j\ Y^2 + \sum_i d_{ij}\ D_i\ Y \tag{17}$$

where $MPC_{ij} = b_j + d_{ij} + 2c_j\ \bar{Y}_i$

$$C_j = a_j + b_j\ Y + c_j\ Y^2 + \sum_i e_{ij}\ D_i\ Y^2 \tag{18}$$

where $MPC_{ij} = b_j + 2(c_j + e_{ij})\ \bar{Y}_i$

$$C_j = a_j + b_j\ Y + c_j\ Y^2 + \sum_i d_{ij}\ D_i\ Y + \sum_i e_{ij}\ D_i\ Y^2 \tag{19}$$

where $MPC_{ij} = b_j + d_{ij} + 2(c_j + e_{ij})\ \bar{Y}_i$

Form (17) allows for shifts in slope between income groups, form (18) allows for shifts in curvature, and form (19) allows for both. It should be noted that all of these expressions still meet the additivity constraint. 1/

Forms (15) to (19) were estimated using OLS regression, and on the basis of conventional statistical criteria, we selected form (19). Since, however, the estimation was done on cross-sectional data which classify expenditures along two criteria (income group and occupation), and since functional form (19) only allows for differences in MPC over income groups, there is an implicit assumption that MPC's do not differ across occupations. We tested that assumption by adding to expression (19) two sets of dummy variables allowing for slope and curvature shifts across k occupation groups. This yields:

$$C_j = a_j + b_j Y + c_j Y^2 + \sum_i d_{ij} D_i Y + \sum_i e_{ij} D_i Y^2 + \sum_k f_{kj} D_k Y + \sum_k g_{kj} D_k Y^2 \qquad (20)$$

The estimation results of (20) show that slope and curvature shifts occur both across income groups and across occupation groups. Consequently, we must compute the MPC_{ij} from expression (20) rather than from (19) in order to avoid misspecification bias. The estimates derived from (20) are best linear unbiased estimates, and at the same time they fulfill the additivity condition since the same functional form is used for all spending categories.

1/ We also estimated functional forms that do not meet the additivity constraint such as the double-log (constant elasticity) Engel function. Equally we experimented with not imposing the requirement that the same form be selected for all expenditure categories. Both cases, however, resulted in a larger violation of the additivity constraint than might be acceptable as compensation for improved statistical goodness of fit resulting from not imposing the constraint.

Expression (20) is a quadratic Engel function with dummy variables allowing slope and curvature shifts both across income groups and across occupation groups. The inclusion of occupational dummy variables guarantees that the estimated coefficients of the income variables pick up the effect of income only, i.e. that they exclude any differences due to occupation. The estimates of equations (20) were the basis for constructing the matrix of income group specific Engel coefficients, the latter being computed at the average income level of each income group.[1]

Let us repeat that the estimation was done on multiple observations _within_ each income group. This is a substantial improvement over the approach necessitated in the case where only a one-way classification is available, viz. to estimate Engel functions (without dummy variables) across income groups, using, for example, functional forms (15) or (16). This approach pre-imposes a continuous function on the data, a continuity for which, we believe, there is no _a priori_ reason. Our methodology allows for discontinuities of the function between income groups, since shifts in slope and curvature are permitted. The use of this methodology depends upon the availability of a two-way classification table as provided by NFIE. This cross-classification is necessary for the estimation of Engel functions (17) to (20) which all contain dummy variables for income groups. A one-way classification across income groups would not leave positive degrees of freedom for the estimation of these functions.

1/ NFIE distinguishes 16 income groups in 1959 and 9 groups in 1969. To make over-time comparisons possible, the results are presented using income quintiles.

III.2.2 The Input-Output Coefficients

The data base used for the Leontief-inverse in our model is the 1960-65-70 Link Input-Output Tables, published by the Administrative Management Agency, Government of Japan (1975). These are a set of consolidated tables, consistent with one another over time, and with the National Accounts. They were specifically prepared "for the time series analysis of economic structure for Japan" (preface). As over time comparability of each block in our model is very important, we chose to use these tables. We used the 1960 input-output table with the 1959 model, and the 1970 table with the 1969 model. Since the NFIE, the data source for the consumption coefficient matrix, uses current prices, we also chose to use the input-output tables in current prices.

The tables come in two formats, 59 x 59 and 159 x 159 sectors. Using the basic table on transactions valued at producers' price, we aggregated the table to 12 x 12 sectors, and then computed the Leontief-inverse $(I - A)^{-1}$. The level of aggregation was determined by the availability of data on the distribution of the wage and profit component of value added (see section III.2.3 of this chapter), which indeed is only available for major sectors. The correspondence between the 12 major sectors and the 59 original ones is listed in Table III.3.

III.2.3. The Value Added Coefficients

The computation of the matrix describing the distribution of households' share of value added over income groups required the combination of several data sources. Information on the share of sectoral value added taken up by wages and profits was found in the 1960-65-70 Link Input-Output Tables. The profit component was broken down into two parts, corporate profits and income of

Table III.3 Correspondence between the 12 x 12 and the 59 x 59 Input-Output Tables

Sector Code	12 x 12 Sectors (abbreviation)	59 x 59 sectors
1	Agriculture (AGR)	agricultural products (1-2) livestock (3-4 forestry, fishery (5-6)
2	Mining (MIN)	coal and lignite metal ore (8-9) petroleum and natural gas (10) other mining (11)
3	Manufacturing (MFG)	food and beverages (12-16) tobacco (17) textiles (18-21) wood products and furniture (22-23) paper products and printing (24-25) leather products (26) rubber products (27) chemicals (28-32) nonmetallic mineral products (33) iron and steel (34-35) metal products (including machinery and equipment (36-42)
4	Construction (CON)	building construction and repairing (43) other construction (44)
5	Electricity, Gas & Water (EGW)	electricity (45) gas (46) water (47)
6	Wholesale & Retail (WRET)	wholesale and retail (48)

Table III.3 (continued) Correspondence between the 12 x 12 and the 59 x 59 Input-Output Tables

Sector Code	12 x 12 Sectors (abbreviation)	59 x 59 Sectors
7	Finance, Insurance & Real Estate (FIR)	finance, insurance (49) real estate (50, 60)
8	Transportation & Communication (TC)	transportation (51) communication (52)
9	Government Service (GOV)	government service (53)
10	Services (SERV)	community services (54) business and personal services (55)
11	Office supplies & (OFF) Packing materials	office supplies (66) packing materials (67)
12	Activities not adequately described (ANAD)	activities not adequately described (57)

Source: The 1960-65-70 Link Input-Output Tables (1975) p. 8-18.

unincorporated enterprises, on the basis of the National Income Accounts. Income of unincorporated enterprises was assumed to go to households completely. We relied upon the results of the Corporation Enterprise Survey to split corporate profits into distributed and retained earnings.

Wages, income of unincorporated enterprises, and distributed corporate profits make up the share of value added going to households. The distribution of that share over income groups was made on the basis of NFIE and FIES tabulations of household income by income group and by sectoral affiliation of the household head. This information is only available for twelve major sectors and determined the level of aggregation of our study. The data also forced us to assume that a person working in a sector obtains his entire income from his productive contributions to that sector. While this is definitionally true for the remuneration of his labor input, it does not necessarily hold for his returns from assets, as one can, of course, hold stocks and bonds from sectors different from the one one works in.[1/] Still, violation of this assumption would not necessarily bias the distributional implications of our results provided the sectoral composition of assets owned would be random with respect to income level, something we unfortunately lack the data to verify.

1/ Weisskoff (1976) made a similar assumption in his study on Puerto Rico. His matrix equivalent to V in our model is said to "summarize the earning and owning practices for each industry." (p.226)

III.2.4. A Note on the Selection of Years

The problem of selecting years is closely tied to the fact that we chose to use marginal rather than average propensities to consume. (The reasons for this choice were explained in section II.4). Indeed, while average propensities could be obtained from the FIES which is a yearly publication, marginal propensities require a cross-tabulation in a second dimension, such as occupation, and at the time of writing, this was available only in the NFIE data for 1959, 1964, 1969 and 1974.

As we wanted to go back as far as possible in Japan's postwar period, the use of the 1959 data became mandatory. We felt that a ten year period was preferable to a five year one in order to observe structural changes with visible impact on income distribution, so the choice for the second[1/] year was between 1969 and 1974. 1974 was an "unusual" year for Japan, as it was suffering from the oil crisis, in the grip of a severe inflationary wage-price spiral, and in the middle of its first severe postwar recession (Patrick and Rosovsky, 1976b, p. 14). While the study of the effects of each of these on income distribution would be very interesting, our model is not designed to cope with them. Consequently, we selected 1969 as our second year. As growth was steady and uninterrupted between 1959 and 1969, the model could clearly perform its function, viz. to capture the effects of the concomitant growth related changes in final demand on distribution.

1/ Time and resource limitations made it necessary to restrict the comparison to two years.

III.3 The Model Results

This section will analyse the solution of the model for each of the selected years. The method of analysis used is comparative statics, i.e. the comparison of the equilibrium solutions for 1959 and 1969 is the basis for explaining what happened between those two years.[1] The purpose is to find out the impact of each final demand component on income distribution. The key of this analysis will be the study of the sectoral composition of each final demand component.

III.3.1 The Model Results for 1959

The discussion of the model results will be based, of course, on the model's solution equation (9) which is reproduced here:

$$\begin{bmatrix} X \\ Y \end{bmatrix} = \begin{bmatrix} B(I - c'VB)^{-1} & Bc'(I - VBc')^{-1} \\ (I - VBc')^{-1}\,VB & (I - VBc')^{-1} \end{bmatrix} \begin{bmatrix} FD - M \\ \bar{Y} \end{bmatrix} \qquad (9)$$

where $B = (I - A)^{-1}$.

1/ Strictly speaking, any over-time comparison of income distribution data should be done using distributions adjusted for changes in age composition and in size of households (Sawyer, 1976, p.18-20). No such adjustments were attempted here. The effects of these changes on the Japanese income distribution have been investigated elsewhere (see the survey in section III.1).

We start the analysis with submatrix $(I - VBc')^{-1}$, describing the linkages between income quintiles. This submatrix for 1959 is presented in Table III.4. The (i,j)th entry of this matrix represents the household income of the ith quintile, induced by expenditures from one unit of income earned by (or transferred to) the jth quintile. All this, of course, after all income-spending-production-income feedbacks have worked themselves out. The entries in the linkages matrix are thus income multipliers. For example, one yen earned by the 3rd quintile or given to it as a government transfer will generate .172949 yen of income for the 2nd quintile.

Looking at the main diagonal, we observe that the within-quintile multipliers increase as one moves to higher quintiles. This means that the spending from a yen earned by the "poor" (or given to them as transfer income) generates less income for the poor themselves than the spending from a yen earned by (or given to) the "rich" generates for the rich. Specifically, a yen earned by or given to the first quintile will ultimately, i.e. in equilibrium, have resulted in receipts of 1.147104 yen for the first quintile. The equivalent multiplier for the fifth quintile is 1.366360, which is 19% higher.

This result is in spite of the fact that the total income generated by a yen originally earned by the poor exceeds that originating from a yen earned by the rich. 1/ This can be seen by looking at the column totals which are total income multipliers. A yen of income of the poorest quintile results in 2.314563 yen of

1/ The same is true, of course, if the yen was not earned by a quintile, but given to it, as a transfer. For the remainder of the discussion, whenever we mention "earned by," the alternative possibility "given to" is implicitly included, and vice versa.

Table III.4. Linkages Matrix, 1959

Income Receiving Quintile	Quintile of Income Origin					
	I	II	III	IV	V	Total
I	1.147104	.150793	.142964	.118925	.083129	1.642913
II	.182554	1.179211	.172949	.151254	.113318	1.799284
III	.207779	.196121	1.193753	.177572	.141620	1.916843
IV	.274649	.257007	.258794	1.243527	.201791	2.235768
V	.502478	.466563	.456943	.440655	1.366360	3.232999
Total	2.314563	2.249693	2.225403	2.131933	1.906217	10.827807

Source: See text.

total household income, while a yen from the richest quintile results in only 1.906217 yen. A policy implication of this is that if income is taken away from the upper quintiles and given to the lower quintiles, total household income will increase, assuming of course, as the model does, that the economy operates under a demand constraint only.

The row totals of the linkages matrix show the total income received by each quintile, if each quintile were to earn (and spend out of) one yen. For example, if each quintile earns and spends out of one yen (i.e. 5 yen in total), quintile I will receive 1.642913 yen or 15% of the 10.827807 yen of total income generated.

As explained in Chapter II, we can interpret the row totals as indicating the minimum level of inequality feasible in the economy, assuming that no changes take place in the consumption, input-output and value added parameters. The row totals of the 1959 linkages matrix for Japan indicate that this minimum corresponds with a Gini-coefficient (G) of .1336 and a Kuznets' measure of inequality (K) of .1313, which is quite a bit below actual inequality of G = .2907 and K = .2747.1/ The policy implication of this is that substantial scope exists to reduce income inequality by exogenous marginal redistribution, for example, by transfer programs, without it being necessary to affect the structure of taxes.

1/ Computational formulas (for quintile distributions) used are

$$\text{Gini-coefficient} = 1 - \sum_i (.2)\,(CYS_i + CYS_{i-1})$$

$$\text{Kuznets-measure} = \frac{\sum_i \left| YS_i - .2 \right|}{1.6}$$

where YS_i = percentage share of income of quintile i

CYS_i = cumulative percentage share of income.

See: Oshima, 1970, p. 8-9; Wada, 1975, p. 229.

The linkages matrix of Table III. 4 is defined as $(I - VBc')^{-1}$ and as such combines the effects of the individual matrices V, B and c' that are its elements. The matrix c', however, combines two effects, viz. the pattern of consumption and the difference in expenditure-income ratio in each quintile. As we are interested in distinguishing the effects of the different consumption patterns alone, we have recomputed the linkages matrix assuming the same expenditure-income ratio for each quintile. For simplicity, we set that ratio equal to one, in other words, we assumed no one saves or pays taxes. The results of that experiment are in Table III. 5.

We notice higher multipliers all over (which is self-evident, since we eliminated some withdrawals from the system), but more interestingly, the differences between the total income multipliers (the row totals) diminished. The consumption pattern of the middle quintile results in the largest amount of household income generated but the differences with the other quintiles are fairly small. This result shows that our previous finding of higher total income multipliers for the lower quintiles is due to a large extent to differences in the expenditure-income ratios between quintiles. The latter are higher for lower income groups, so less is withdrawn from the multiplier process.

The income distribution implied by the row totals corresponds with G = .1742 and K = 1712. The fact that endogenous inequality is higher when it is assumed that there are no saving or taxes, implies that the existing pattern of saving and taxes is beneficial with respect to income distribution. 1/ This is confirmed by

---

1/ This is only true for a given distribution of wealth. This model does not consider the effects of different saving rates on wealth accumulation. The fact that higher income groups have a higher average propensity to save allows them to accumulate wealth faster and could thus change the distribution of income in their favor. This would be reflected by a change in V in the linkages matrix $(I-VBc')^{-1}$.

Table III.5 Linkages Matrix, 1959, Assuming No Saving or Taxes

Income Receiving Quintile	Quintile of Income Origin					
	I	II	III	IV	V	Total
I	1.210404	.235994	.241137	.219799	.199105	2.106436
II	.265254	1.288245	.298401	.282027	.267645	2.401570
III	.307385	.324928	1.341905	.333950	.330565	2.638733
IV	.413543	.434545	.463108	1.460606	.467863	3.239663
V	.753697	.787863	.824178	.833315	1.848373	5.047424
Total	2.950280	3.071573	3.168727	3.129695	3.113550	15.433826

Source: See text.

the fact that setting the expenditure-income ratios equal to one for all quintiles barely increases the within-quintile multiplier of the poorest quintile but it increases the one of the richest quintile substantially. In other words, without taxes or saving, the feedback from the spending of the rich to their own income would be much higher.

The eigenvector associated with the 1959 linkages matrix and representing the steady state distribution corresponds with a G = .2574 and K = .2249 which is a bit below actual inequality (G = .2907, K = .2747) but well above the minimum endogenous inequality (G = .1336, K= .1313). The latter, it will be recalled, we interpreted as the distribution which would result from an exogenously implemented (re)distribution that is perfectly equal at the margin. The level of endogenous inequality cannot sustain itself but would deteriorate to the level of inequality indicated by the eigenvector. The policy implication is that, in the long run, equality can only be brought below the level implicit in the linkages matrix eigenvector by continuous exogenous redistributive efforts.

The next step in our analysis is to see how each sector in the economy affects income generation and distribution. For that purpose, we look at submatrix $(I - VBc')^{-1}$ VB of solution equation (9). It is the product of the linkages matrix with the matrix VB, and describes the effects of a one yen increase of final demand in each sector on the incomes of each quintile. This matrix holds the key to explaining how shifts in the sectoral composition of final demand affect income distribution. Since it gives directly the equilibrium implications for household income and its distribution, this matrix would be useful for project analysis in developing countries.

The matrix $(I - VBc')^{-1}$ VB for 1959 is presented in Table III.6. The activities with the highest multipliers are agriculture, government spending and services. A one yen increase of final demand for agricultural output, for example, will generate 1.746893 yen of household income. Agriculture and services are sectors with a large percentage of unincorporated enterprises, which, in part, explains the high multipliers. In the case of government spending it is the large percentage of value added going to wages that accounts for the high multiplier.

Activities particularly favorable for the poorest income group are agriculture and construction. [1/] One yen of spending in those sectors generates respectively .271850 and .197823 yen of income for the first quintile. Particularly low first quintile multipliers are found in the finance-insurance-real estate sector, and in the utilities sector, viz. .060194 and .095795, respectively.

The richest quintile derives, on a per unit basis, the most income from final demand for services, wholesale and retail, and agriculture. Each of these sectors generates, for each yen of final demand, about .60 yen for the richest quintile.

The activity with the highest multiplier for the middle groups is government spending.

In terms of the summary measures G and K reported in Table III.6, the highest levels of inequality are generated by final demand in finance, insurance and real estate, and in services. The lowest levels of inequality stem from final demand in agriculture and construction.[2/]

1/ This is so looking only at the multipliers, and disregarding the absolute amounts of final demand that go to these sectors.

2/ We should emphasize that the inequality of income we talk about refers to incomes generated by final demand in the sectors in question. This is not to be confused with the inequality of incomes paid to people who work in a given sector. For example, the distribution of income of construction workers is not the same as the distribution of incomes generated by final demand for construction. The latter includes incomes in other sectors that deliver to construction.

Table III.6 Income Generation Per Sector, 1959

Quintile	AGR	MIN	MFG	CON	EGW	WRET
I	.271850	.159324	.153532	.197823	.095795	.151318
II	.287460	.215634	.185918	.225087	.136039	.198913
III	.272583	.267022	.211443	.235234	.168666	.222730
IV	.327358	.309035	.278266	.296319	.211696	.299189
V	.587642	.486655	.477318	.521600	.339204	.591845
Total	1.746893	1.437670	1.306477	1.476063	.951397	1.463995
G	.1538	.2081	.2265	.1948	.2365	.2681
K	.1705	.1918	.2229	.1926	.2238	.2608
	FIR	TC	GOV	SERV	OFF	ANAD
I	.060194	.120002	.158267	.146799	.153106	.165385
II	.090008	.175816	.250898	.194438	.188325	.185130
III	.113908	.220610	.299115	.235836	.213720	.207787
IV	.159500	.282125	.364469	.335998	.281976	.267280
V	.396005	.450011	.508098	.623319	.499030	.468182
Total	.819615	1.248564	1.580847	1.536390	1.336157	1.293764
G	.3617	.2455	.2058	.2850	.2351	.2126
K	.3539	.2330	.1899	.2805	.2306	.2106

Source: See text.

We are now in a position to investigate the contribution of each final demand component to the process of income generation and distribution. The final demand data by component were aggregated from the 1960-65-70 Link Input-Output Tables and are presented in Table III.7. The incomes generated by each final demand component are in Table III.8, both in absolute figures and in percentages. These tables highlight the advantage of using a model which breaks down final demand into its components and which explicitly recognizes the sectoral composition of each component. These two features indeed make possible the computation of the amount of household income, and its distribution, generated by each final demand component separately. This yields more information on the factors that contribute to an overall observed distribution than could be obtained from any formal decomposition of inequality summary measures.

The most important final demand component in absolute terms is investment. It is also the component which generates the most equal income distribution ($G = .2168$, $K = .2122$). This is predominantly due to the importance of the construction sector in investment, a sector which generates a very low level of inequality. The second most important component is government spending which channels income mostly into the middle quintiles.

It is interesting to note that the level of inequality associated with exports is higher than that which would result had imports been produced domestically. Again, the explanation is given by the sectoral composition: major export sectors are manufacturing and transportation and communication (more than 90% of total exports), which both have higher inequality associated with them than agriculture and mining, which make up more than 50% of total imports.

Table III.7. Autonomous Final Demand, 1959

	C_o		G		I		Inv		X		X_s		M	
	10^6yen	%	10^6yen	%	10^6yen	%	10^6yen	%	10^6yen	%	10^6yen	%	10^6yen	%
AGR	6999	0.79	935	0.06	16149	0.35	81880	12.91	56631	3.28	1400	1.18	563671	31.87
MIN	0	0	1638	0.10	0	0	1073	0.17	856	0.05	11	0.01	387521	21.91
MFG	221574	25.03	109477	6.57	1566198	33.62	516279	81.42	1331336	77.23	41842	35.12	702750	39.73
CON	0	0	6225	0.37	2876640	61.75	0	0	132	0.01	5753	4.83	171	0.01
EGW	192	0.02	26066	1.56	0	0	1	0.00	759	0.04	3949	3.31	794	0.05
WRET	133093	15.04	13640	0.82	158330	3.40	22902	3.61	61086	3.54	1122	0.94	13497	0.76
FIR	0	0	47924	2.87	0	0	0	0	2268	0.13	0	0	0	0
TC	10385	1.17	50111	3.01	14196	0.30	8108	1.28	244503	14.18	7216	6.06	58829	3.33
GOV	0	0	818286	49.08	0	0	0	0	0	0	0	0	0	0
SERV	342117	38.65	591679	35.49	0	0	0	0	3367	0.20	341	0.29	5895	0.33
OFF	0	0	4642	0.28	0	0	0	0	0	0	0	0	0	0
ANAD	170741	19.29	-3359	-0.20	27112	0.58	3858	0.61	22991	1.33	57497	48.26	35785	2.02
TOTAL	885101	100.00	1667264	100.00	4658625	100.00	634101	100.00	1723929	100.00	119131	100.00	1768913	100.00

Source: 1960-65-70 Link Input-Output Table, p. 34-35.

Code:
C_o - consumption expenditures outside household
G - government consumption expenditures
I - gross domestic fixed capital formation
I_{nv} - increase in stocks

X - exports of goods and services (except special procurement export)
X_s - special procurement exports (deliveries to U.S. military forces stationed in Japan)
M - imports of goods and services

Table III.8. Income Generated by Final Demand Component , 1959

Quintile	C_o	G	I	I_{nv}	X	X_s	M
	(a) 10^6 yen						
I	135786	248535	844063	106772	263050	18918	338865
II	169662	362363	982330	126450	324369	22219	397195
III	196887	433848	1056277	139461	370701	25120	430700
IV	267324	561955	1352131	180969	484363	32499	532231
V	486569	900610	2370308	314077	829436	56183	910510
Total	1256227	2507311	6605108	867729	2271919	154940	2609499
	(b) percentages						
I	10.81	9.91	12.80	12.30	11.58	12.21	12.98
II	13.51	14.45	14.87	14.57	14.28	14.34	15.22
III	15.67	17.30	15.99	16.07	16.32	16.21	16.50
IV	21.28	22.41	20.47	20.86	21.32	20.98	20.40
V	38.73	35.92	35.89	36.19	36.51	36.26	34.89
Total	100.00	100.00	100.00	100.00	100.00	100.00	100.00
G	.2545	.2399	.2072	.2163	.2276	.2189	.1959
K	.2502	.2291	.2045	.2131	.2228	.2155	.1911

Source: See text.

From a policy point of view, it is important to know not only the distributional impact of spending, but also how much income is generated per unit of spending. For that purpose we computed multipliers per unit of final demand, reported in Table III.9. 1/ A yen of government spending generates more household income (1.503848 yen) than any other final demand component. Virtually tied for second place are investment and consumption outside households (multipliers are 1.417823 and 1.419303, respectively).

Comparing government spending with investment, we notice (from Table III.8 (b)) that the richest quintile receives about 35% of the total income generated by either. On a per unit basis, however, government spending is more favorable to the richest quintile than is investment.

From Table III.8 (b) we also notice that, percentage-wise, investment is more favorable for the poorest income group. Of all household incomes generated by investment, 12.80% flows to the poorest quintile versus only 9.91% of all household incomes generated by government spending. The multipliers of Table III.9 fortify that conclusion by pointing out that also on a per unit basis, investment spending results in more income for the poorest quintile, and this is in spite of the overall higher multiplier of government spending.

Altogether, we notice the very special role of investment. It generates the most equal income distribution, and at the same time, it has the second highest household income multiplier per unit of spending. It is also in absolute size the largest component of exogenous final demand. As such we may conclude that investment spending was not only a major force beyond the level (and growth) of

1/ The entries in that table are obtained by dividing the entries of Table III.8 (a), total income generated by each final demand component, by the corresponding column totals of Table III.7 (amounts of total final demand).

Table III.9. Multipliers Per Unit of Final Demand, 1959

Quintile	C_o	G	I	I_{nv}	X	X_s	M
I	.153413	.149067	.181183	.168383	.152587	.158800	.191567
II	.191686	.217340	.210863	.199416	.188157	.186509	.224542
III	.222446	.260215	.226736	.219935	.215033	.210860	.243483
IV	.302026	.337052	.290242	.285394	.280964	.272800	.300880
V	.549733	.540172	.508800	.495311	.481131	.471607	.514728
Total	1.419303	1.503848	1.417823	1.368440	1.317873	1.300585	1.475199

Source: See text.

household incomes, but, at the same time, a major explanatory factor of the fairly low level of inequality in the distribution of household income in 1959. 1/

The above argument can further be illustrated by Table III.10 which lists the relative importance of each final demand component in the income of each quintile. The role of investment clearly stands out, as it provides, on the average, 48.34% of all household incomes. The poorest quintile, however, received 52.20% of its income from investment. Government spending generates, on the average, 18.35% of all household incomes, but a higher percentage of the incomes of the middle quintiles.

We will conclude the discussion of the 1959 model solution with an experiment. Suppose we set vector $\bar{Y}$ in solution equation (9) equal to the existing income distribution (with FD - M set equal to zero), and solve to obtain Y, the equilibrium income distribution corresponding to it. In other words, we obtain Y from

$$\begin{bmatrix} X \\ \hline Y \end{bmatrix} = \left[\begin{array}{c|c} B(I - c'VB)^{-1} & Bc'(I - VBc')^{-1} \\ \hline (I - VBc')^{-1}\, VB & (I - VBc')^{-1} \end{array}\right] \begin{bmatrix} 0 \\ \hline \bar{Y} \end{bmatrix}$$

where $\bar{Y}$ is the existing income distribution.

1/ This conclusion does not run counter to what one might expect a priori. Indeed, investment involves a lot of construction which is a sector with great technical flexibility. It offers much opportunity for labor-intensive methods of production, and as such it can be favorable for the lower income groups. Ranis (1971) and Turnham and Hawkins (1973) both discuss the importance of the construction sector in providing employment and in channeling income to low income groups in LDC's. Although Japan could hardly be classified as LDC in 1959, its experience seems quite relevant for today's LDC's in clearly bringing out the importance of the construction sector as a provider of income for the poor.

Table III.10 Relative Importance of Final Demand Components as Income Source, 1959

	C_o	G	I	I_{nv}	X	X_s	Total
I	8.40	15.37	52.20	6.60	16.27	1.17	100.00
II	8.54	18.23	49.43	6.36	16.32	1.12	100.00
III	8.86	19.52	47.53	6.38	16.68	1.13	100.00
IV	9.28	19.52	46.96	6.29	16.82	1.13	100.00
V	9.81	18.17	47.82	6.34	16.73	1.13	100.00
Average	9.19	18.35	48.34	6.35	16.63	1.13	100.00

Source: Table III.8 (each entry is obtained by dividing each entry of Table III.8 by its corresponding row total, excluding imports).

The result presented in Table III.11 shows that the existing income distribution is virtually an equilibrium one (although the latter yields much higher income levels). The result is consistent with our interpretation of the row totals of the linkages matrix as portraying the level of endogenous inequality. Indeed as the latter is below the level of the existing distribution, the equilibrium distribution is slightly more equal than the actual one.

III.3.2 The Model Results for 1969

We shall discuss the 1969 model fully analogously to the 1959 model.

Compared to 1959, all multipliers in the 1969 linkages matrix (Table III.12) are smaller. Otherwise, the same conclusions can be derived from the 1969 linkages matrix as from the 1959 one, viz. the within-quintile multipliers increase as one moves to higher quintiles, and the total income multiplier for income originally earned by the poor exceeds the one for the income of the rich. With respect to the latter, the difference increased: in 1959 the total income multiplier of income originally earned by the fifth quintile was 18% below the multiplier for income earned by the first quintile, in 1969 it is 28% lower.

The endogenous inequality implied by the 1969 linkages matrix corresponds with G = .1301 and K =.1231, a slight decrease relative to 1959 (for 1959, G =.1336, K = .1313). Since we know that observed total inequality increased between the two years, we can already presume that the explanation will have to be found in changes that took place in the composition of final demand.

In 1969 the eigenvector of the linkages matrix reflecting the steady state income distribution corresponded with G = .2778 and K = .2287. This is a bit higher than in 1959. As in 1959, however, it is only a little below actual but well above endogenous inequality.

Table III.11 Equilibrium Income Distribution, 1959

	(1)	(2)	(3)	(4)
I	13204	8.06	32074	9.38
II	21270	12.98	45190	13.21
III	27825	16.98	55819	16.32
IV	36349	22.18	74651	21.82
V	65196	39.79	134372	39.28
G	.2907		.2737	
K	.2747		.2637	

Source: See text.

(1) original income distribution as computed from 1959 NFIE, Table 1-1, p. 31-32; average monthly income in yen.

(2) column (1) expressed as percentages.

(3) <u>e</u>quilibrium income distribution where $\overline{Y}$ = col. (1); average monthly income in yen.

(4) column (3) expressed as percentages.

Table III.12 Linkages Matrix, 1969

Income Receiving Quintile	Quintile of Income Origin					
	I	II	III	IV	V	Total
I	1.091853	.081553	.074269	.067670	.044192	1.359535
II	.152028	1.133187	.121339	.111393	.071718	1.589663
III	.201928	.177373	1.161745	.147728	.095369	1.784142
IV	.263469	.233351	.213252	1.194310	.126270	2.030650
V	.423460	.378647	.348614	.315848	1.207425	2.673994
Total	2.132736	2.004109	1.919218	1.836948	1.544973	9.437984

Source: See text.

As the following table shows, changes occurred in the expenditure-income ratios for each quintile between 1959 and 1969.

Expenditure-Income Ratios

Quintile	1959	1969
I	.995	1.012
II	.886	.877
III	.831	.824
IV	.788	.761
V	.641	.489

Consequently, we will also want to compare the linkages matrices computed under the assumption that the ratios equal one for all quintiles. This matrix for 1969 is represented in Table III.13.

We note again, of course, that all multipliers are smaller than in 1959, but the decrease is not the same for all quintiles. Table III.13 (b) expresses the decrease in the value of the multipliers as a percentage of their 1959 value. We notice that the multipliers expressing the income received by the third and fourth quintiles fell considerably less than the others. This suggests that, in terms of deriving income from the spending of other groups, the middle class in Japan strengthened its relative position between 1959 and 1969.

The (endogenous) income distribution implied by the matrix corresponds with G = .1736 and K = .1644, values slightly lower than in 1959, but higher than the ones obtained from the linkages matrix computed with actual expenditure-income

Table III.13 Linkages Matrix, 1969, Assuming No Saving or Taxes

Income Receiving Quintile	Quintile of Income Origin					
	I	II	III	IV	V	Total
(a) Linkages Matrix						
I	1.135046	.138398	.134571	.132529	.134771	1.675313
II	.222354	1.225826	.219637	.217403	.218975	2.104194
III	.295434	.300584	1.292529	.288557	.291173	2.468276
IV	.387063	.396015	.385967	1.380118	.385267	2.934429
V	.625949	.644542	.631330	.619395	1.632243	4.153458
Total	2.665843	2.705365	2.664034	2.638001	2.662428	13.335670
(b) Percentage Change Between 1959 and 1969						
I	- 6.23	-41.36	-44.19	-39.70	-32.31	-20.47
II	-16.17	- 4.84	-26.40	-22.91	-18.18	-12.38
III	- 3.89	- 7.49	- 3.68	-13.59	-11.92	- 6.46
IV	- 6.40	- 8.87	-16.66	- 5.51	-17.65	- 9.42
V	-16.95	-18.19	-23.40	-25.67	-11.69	-17.71
Total	- 9.64	-11.92	-15.93	-15.71	-14.49	-13.59

Source: See text.

ratios. Again (see the discussion in section III.3.1 of this chapter), this means that the existing pattern of saving and taxes is favorable for equality in the distribution of incomes (excluding any wealth effects).

Table III.14 differentiates the income generation process per sector. Agriculture, government spending and services are, as in 1959, the sectors with the highest total household income multipliers.

Wholesale and retail, and services joined agriculture and construction as activities with high multipliers for the lowest quintile, ranging between .104745 and .114791. Low first quintile multipliers are again found in the finance-insurance-real estate sector and in the utilities sector (respectively .041492 and .050135).

The three sectors with the highest overall multiplier together with wholesale and retail have also the highest multipliers for the top quintile, ranging from .464174 to .553881.

As in 1959, the government sector displays the highest multiplier for the middle groups.

Overall inequality, as reported by the summary measures in Table III.14 rose in almost all sectors, but especially in agriculture, government spending and utilities. Construction and mining are now the sectors resulting in the lowest level of inequality.

We can now investigate the constribution of each final demand component to the process of income generation and distribution. The figures are in Tables III.15 and III.16.

As in 1959, investment still generates, by far, the most household income. Its distribution, however, has become less equal as the share of the poorest

Table III.14. Income Generation Per Sector, 1969

Quintile	AGR	MIN	MFG	CON	EGW	WRET
I	.114791	.095702	.083732	.109470	.050135	.112754
II	.174174	.174452	.142244	.172526	.094890	.164345
III	.236335	.205978	.186451	.202021	.144868	.200151
IV	.321490	.257649	.243406	.249774	.216517	.255673
V	.553881	.309941	.387798	.414513	.303429	.472057
Total	1.400671	1.043722	1.043661	1.148304	.809839	1.204980
G	.2929	.1961	.2719	.2394	.3103	.2689
K	.2812	.1798	.2560	.2231	.3025	.2549
	FIR	TC	GOV	SERV	OFF	ANAD
I	.041492	.079492	.072602	.104745	.089988	.121137
II	.072790	.156943	.145180	.158370	.146001	.132003
III	.099362	.229165	.237660	.213120	.188588	.154654
IV	.144610	.266490	.344540	.288315	.244484	.200784
V	.276761	.377502	.464174	.493137	.399094	.304267
Total	.635015	1.109592	1.264156	1.257687	1.068155	.912845
G	.3416	.2543	.3109	.2884	.2684	.1906
K	.3294	.2336	.2997	.2767	.2531	.1916

Source: See text.

Table III.15. Autonomous Final Demand, 1969

	C_o		G		I		I_{nv}		X		X_s		M	
	10^6yen	%	10^6yen	%	10^6yen	%	10^6yen	%	10^6yen	%	10^6yen	%	10^6yen	%
AGR	37063	1.17	52	0.00	94043	0.36	4991	0.21	87125	1.04	2918	1.68	1638047	21.20
MIN	0	0	942	0.02	0	0	58232	2.48	4798	0.06	1302	0.75	2146099	27.78
MFG	743561	23.50	418830	6.70	9479802	36.10	2103941	89.17	6456675	77.13	59953	34.48	3072915	39.78
CON	0	0	44332	0.71	14871122	56.63	0	0	0	0	0	0	0	0
ECW	337	0.01	177189	2.84	0	0	0	0	68	0.00	3449	1.98	51	0.00
WRET	274721	8.68	47267	0.76	1722041	6.56	151566	6.42	444810	5.31	12017	6.91	119160	1.54
FIR	0	0	298315	4.77	0	0	0	0	24840	0.30	386	0.22	29880	0.39
TC	15483	0.49	175860	2.81	109347	0.42	18459	0.78	1148246	13.72	41716	23.99	402102	5.20
GOV	0	0	2313152	37.02	0	0	0	0	0	0	0	0	0	0
SERV	2028137	64.10	2487570	39.81	0	0	0	0	39147	0.47	4107	2.36	93790	1.21
OFF	0	0	25092	0.40	0	0	0	0	0	0	0	0	0	0
ANAD	64542	2.04	259967	4.16	-18446	-0.07	22392	0.95	165413	1.98	48032	27.62	223290	2.89
Total	3163844	100.00	6248568	100.00	26257909	100.00	2359581	100.00	8371127	100.00	173880	100.00	7725334	100.00

Source: 1960-65-70 Link Input-Output Tables, p. 66-67

Code: See Table III.7

Table III.16. Income Generated by Final Demand Component , 1969

Quintile	C_o	G	I	I_{nv}	X	X_s	M
	(a) 10^6 yen						
I	318994	542838	2633125	203583	717694	16588	734235
II	489549	909054	4228215	341063	1217597	25131	1225999
III	648198	1317157	5160897	443485	1614003	32943	1575620
IV	865042	1826241	6517810	576886	2068668	41687	2041350
V	1464298	2795676	10741178	922046	3273602	64483	3094638
Total	3786079	7390963	29281200	2487062	8891561	180832	8671841
	(b) percentage						
I	8.42	7.34	8.99	8.19	8.07	9.17	8.47
II	12.93	12.30	14.44	13.71	13.69	13.90	14.14
III	17.12	17.82	17.62	17.83	18.15	18.22	18.17
IV	22.85	24.71	22.26	23.19	23.27	23.05	23.54
V	38.68	37.83	36.68	37.07	36.82	35.66	35.69
Total	100.00	100.00	100.00	100.00	100.00	100.00	100.00
G	.2817	.2935	.2528	.2690	.2682	.2485	.2554
K	.2690	.2817	.2368	.2534	.2510	.2339	.2403

Source: See text.

quintile is reduced by almost four percentage points in favor of the middle quintiles. This phenomenon is typical, in fact, for all other final demand components. As a result, the level of inequality associated with each final demand component went up.

This can be traced back, to a large extent, to the altered sectoral composition of each component. For example, in 1969, the share of the construction sector in investment fell compared to 1959, while the shares of the manufacturing and wholesale and retail sectors rose. The sectoral composition of government spending also shifted between 1959 and 1960 toward more spending in the services, utilities and finance-insurance-real estate sectors, which are all characterized by fairly high inequality of the incomes they generate.

Note that income from exports is again, as in 1959, distributed less equally than income from imports, had they been produced domestically.

The multipliers per unit of final demand reported in Table III.17 are lower in 1969 than in 1959. This is in line with the changes observed in the linkages matrix. Contrary to 1959, consumption outside the household has the highest multiplier. This is due to a large shift to services in that component. Government spending comes in second. Investment is only third, but still has the highest multiplier for the poorest quintile (a position it shares though with consumption outside households).

In addition, looking at Table III.18 the relative importance of investment as income source increased for all quintiles, while especially government spending and consumption outside the household became less important.

We believe, therefore, that the increased importance of investment as a source of income, combined with a less equal distribution of that income, is a major explanatory factor in the observed increase in income inequality in Japan between

Table III.17. Multipliers Per Unit of Final Demand, 1969

Quintile	C_o	G	I	I_{nv}	X	X_s	M
I	.100825	.086874	.100279	.086279	.085734	.095399	.095042
II	.154732	.145482	.161026	.144544	.145452	.144531	.158698
III	.204877	.210793	.196546	.187951	.192806	.189458	.203955
IV	.273415	.292265	.248223	.244487	.247119	.239746	.264241
V	.462822	.447411	.409064	.390767	.391059	.370848	.400583
Total	1.196670	1.182825	1.115138	1.054027	1.062171	1.039981	1.122520

Source: See text.

Table III.18. Relative Importance of Final Demand Components as Income Source, 1969

	C_o	G	I	I_{nv}	X	X_s	Total
I	7.20	12.25	59.40	4.59	16.19	.37	100.00
II	6.79	12.61	58.64	4.73	16.89	.35	100.00
III	7.03	14.29	56.00	4.81	17.51	.36	100.00
IV	7.27	15.35	54.79	4.85	17.39	.35	100.00
V	7.60	14.51	55.77	4.79	17.00	.34	100.00
Average	7.28	14.21	56.29	4.78	17.09	.35	100.00

Source: Table III.16.

1959 and 1969. However, had government spending kept its 1959 relative importance, the total income distribution would have been more unequal, since the increase in inequality was greater in the incomes derived from government spending than in those derived from investment.

Finally, Table III.19 shows that also in 1969 the existing income distribution is virtually an equilibrium one (although the latter has again higher income levels).

III.4 Summary

We started this chapter with a review of studies on Japanese income distribution which revealed that they had concentrated on the impact on distribution of demographic and institutional changes and of changes in the labor market. An application of the demand oriented model developed in Chapter II to the Japanese case appeared therefore warranted. This chapter described how the model parameters were estimated from Japanese data and investigated the solution of the model for 1959 and 1969.

We examined first the linkages matrix which describes how the processes of consumption, production, and income generation link different income groups together economically.

We found that higher income groups have a higher within-group multiplier, i.e. their spending tends to generate more income for themselves than is the case for the lower income groups. This phenomenon is mitigated though by the fact that higher income groups have lower expenditure-income ratios, which means more withdrawals from the multiplier process.

Table III.19 Equilibrium Income Distribution, 1969

	(1)	(2)	(3)	(4)
I	41949	8.48	73014	8.31
II	63942	12.92	114759	13.06
III	80276	16.22	147850	16.83
IV	101903	20.59	190909	21.73
V	206806	41.79	351849	40.06
G	.2972		.2886	
K	.2798		.2724	

Source: See text.

(1) original income distribution as in 1969 NFIE, Table 2-1, p. 56-57; average monthly income in yen.

(2) column (1) expressed as percentages.

(3) equilibrium income distribution where $\overline{Y}$ = column (1); average monthly income in yen.

(4) column (3) expressed as percentages.

On a per unit basis, income earned and spent by higher income groups generates less total household income (for all income groups combined), than income originally earned by (or transferred to) poor groups. This is due to a large extent to the higher expenditure-income ratios of the poor groups.

Between 1959 and 1969 the structure of income linkages changed. It appears that the third and fourth quintiles strengthened their relative position, in that in 1969 they derive a larger share of the income generated by the spending of other groups than in 1959.

The linkages matrix indicates the minimum level of inequality in income distribution feasible, given certain structural elements of the economy. In 1959 this "endogenous inequality" in Japan corresponded with a Gini-coefficient (G) as low as .1336. Between 1959 and 1969, endogenous inequality slightly fell to G = .1301.

The existing pattern of saving and taxes was found to be an equalizing factor with respect to the level of endogenous inequality, at least when wealth effects are excluded (see footnote 1, p.48).

With the linkages matrix is associated an eigenvector which reflects the steady state income distribution of the economy. Both in 1959 and 1969 the inequality of this distribution was slightly less than that of the actual distribution, but substantially higher than endogenous inequality. The comparison of endogenous and steady state levels of inequality with one another and with actual inequality has important policy implication, for example, with respect to the choice of type of policy programs to equalize income distribution.

Since during the same time period that endogenous inequality decreased, the distribution of total household income became less equal, the explanation needs to be found in changes in the composition of final demand.

We found that final demand in different sectors is associated with widely different levels of income inequality of the incomes generated. In 1959, high levels of income inequality were generated by final demand in finance, insurance and real estate sector, and in services. The lowest level of inequality stemmed from final demand in agriculture and construction. In 1959, finance, insurance and real estate, the government sector, and the utilities sector generated the least equally distributed income. Construction and mining were the sectors yielding the lowest levels of inequality.

Of all exogenous final demand components, investment generates the most equal income distribution, because of the importance of the construction sector for investment. Since investment is the largest component of exogenous final demand, it plays a major role in explaining the low level of inequality in the distribution of household incomes in 1959. The increased importance of investment as a source of income in 1969, combined with a less equal distribution of that income, explains in part the observed increase in income inequality between 1959 and 1969. At the same time it helped to check that increase, which indeed would have been larger had government spending (the second largest component of exogenous final demand) kept its 1959 relative importance, since the increase in inequality was greater in the incomes derived from government spending than in those derived from investment.

CHAPTER IV

A SIMULATION EXERCISE

IV.1 Simulation Experiments

The previous chapter investigated the relationship between final demand and income distribution, as it was captured by our model, in the years 1959 and 1969. At several points a comparison between the two years was made, mostly, however, without explicitly investigating the causes of the observed changes. The latter is the topic of this chapter.

This chapter will investigate the relative importance of the changes that took place between 1959 and 1969 in the pattern of consumption, the input-output structure, the value added distribution, and the composition of exogenous final demand, in explaining the changes in income distribution that were observed between 1959 and 1969.

The technique of analysis is a series of simulation exercises whereby, one by one, the elements of the 1969 model are substituted for the corresponding 1959 elements and each time the resulting hypothetical change in income distribution is interpreted as the relative contribution of that model element to the observed changes in income distribution. 1/

We conducted 14 simulation experiments, listed in Table IV.1. The major results are presented in Table IV.2, which shows the summary inequality measures of the equilibrium income distribution implied by exogenous final demand, the total amount of household income generated, and the implied multiplier (obtained by dividing total household income by exogenous final demand).

1/ The same technique was used by Weisskoff (1976); see his footnote 12 (p. 209) for additional references.

Table IV.1. Simulation Experiments

#1	:	V (59)	B (59)	C (59)	FD (69)
#2	:	V (59)	B (69)	C (59)	FD (59)
#3	:	V (59)	B (69)	C (59)	FD (69)
#4	:	V (59)	B (59)	C (69)	FD (59)
#5	:	V (59)	B (59)	C (69)	FD (69)
#6	:	V (69)	B (59)	C (59)	FD (59)
#7	:	V (69)	B (59)	C (59)	FD (69)
#8	:	V (59)	B (69)	C (69)	FD (59)
#9	:	V (59)	B (69)	C (69)	FD (69)
#10	:	V (69)	B (69)	C (59)	FD (59)
#11	:	V (69)	B (69)	C (59)	FD (69)
#12	:	V (69)	B (59)	C (69)	FD (59)
#13	:	V (69)	B (59)	C (69)	FD (69)
#14	:	V (69)	B (69)	C (69)	FD (59)

where: V = distribution matrix of households' share of value added

B = Leontief-inverse matrix

C = matrix of consumption coefficients

FD = matrix of final demand structure

Table IV.2. Results of the Simulations

Experi-ment #	Gini-Coefficient	Kuznets Measure	Total Income Generated (10^6 yen)	Income Multiplier
1959	.2277	.2230	11053735	1.395808
1	.2278	.2238	54176336	1.394515
2	.2463	.2403	9163800	1.157157
3	.2468	.2417	44980064	1.157801
4	.2111	.2066	9969623	1.258912
5	.2111	.2074	48859728	1.257664
6	.2727	.2585	11826889	1.493438
7	.2725	.2581	57975216	1.492300
8	.2318	.2258	8191678	1.034402
9	.2322	.2271	40202800	1.034832
10	.2703	.2559	9802108	1.237759
11	.2700	.2554	48154896	1.239522
12	.2684	.2537	10690595	1.349952
13	.2682	.2532	52401104	1.348820
14	.2661	.2510	8823972	1.114245
1969	.2658	.2504	43345872	1.115736

Source: See text.

IV.2 The Results

The first experiment substitutes the 1969 final demand for the 1959 one, while otherwise keeping all 1959 model elements. Inequality of total household income seems hardly affected. Both G and K are increased, but the change seems insignificant. This is a surprising result for two reasons.

First, it is not a reflection of the fact that only minor or no changes took place in the composition of final demand. In order to illustrate this, we have reproduced in Table IV.3 the sectoral composition of each final demand component in 1959 and 1969. We note some quite significant sectoral shifts within all of the final demand components.

Consumption outside households saw an enormous shift to services, away from wholesale and retail, and from the miscellaneous sector (ANAD). Consumption outside households includes mainly social and travel expenses by businessmen and the observed sectoral shift thus indicates an important increase in the service component of expense accounts. The spending pattern of the government sector shifted away, perhaps strangely, from government services itself toward services, finance, insurance and real estate, utilities, and the miscellaneous sectors. Investment became less construction, more manufacturing (i.e. machines and equipment) and retail and wholesale. Additions to inventories in agriculture dropped to virtually zero in 1969, while more of the increases in inventories occurred in the manufacturing and wholesale and retail sectors. Exports was the most stable final demand component with only a small shift away from agriculture to wholesale and retail. Imports shifted away from agriculture to mining products (undoubtedly oil) and transportation and communication.

Table IV.3. Autonomous Final Demand by Category, 1959 and 1969

	C_o		G		I		I_{nv}		X		X_s		M	
	59	69	59	69	59	69	59	69	59	69	59	69	59	69
AGR	0.79	1.17	0.06	0.00	0.35	0.36	12.91	0.21	3.28	1.04	1.18	1.68	31.87	21.20
MIN	0	0	0.10	0.02	0	0	0.17	2.47	0.05	0.06	0.01	0.75	21.91	27.78
MFG	25.03	23.50	6.57	6.70	33.62	36.10	81.42	89.17	77.23	77.13	35.12	34.48	39.73	39.78
CON	0	0	0.37	0.71	61.75	56.63	0	0	0.01	0	4.83	0	0.01	0
EGW	0.02	0.01	1.56	2.84	0	0	0.00	0	0.04	0.00	3.31	1.98	0.05	0.00
WRET	15.04	8.68	0.82	0.76	3.40	6.56	3.61	6.42	3.54	5.31	0.94	6.91	0.76	1.54
FIR	0	0	2.87	4.77	0	0	0	0	0.13	0.30	0	0.22	0	0.39
TC	1.17	0.49	3.01	2.81	0.30	0.42	1.28	0.78	14.18	13.72	6.06	23.99	3.33	5.20
COV	0	0	49.08	37.02	0	0	0	0	0	0	0	0	0	0
SERV	38.65	64.10	35.49	39.81	0	0	0	0	0.20	0.47	0.29	2.36	0.33	1.21
OFF	0	0	0.28	0.40	0	0	0	0	0	0	0	0	0	0
ANAD	19.29	2.04	-0.20	4.16	0.58	-0.07	0.61	0.95	1.33	1.98	48.26	27.62	2.02	2.89
Total	100.00	100.00	100.00	100.00	100.00	100.00	100.00	100.00	100.00	100.00	100.C0	100.00	100.00	100.00

Source: Tables III. 7 and III.15.

A second reason why the seeming distributional neutrality of the changes in final demand composition is surprising is because, as we found in Chapter III (see Tables III.6 and III.14), the distributional implications of a yen spent in each (input-output) sector vary greatly across sectors. For example, in 1959, income spent on agricultural items generated income that was very evenly distributed (G = .1538, K = .1705), while income spent in the finance, insurance and real estate sector resulted in very unequally distributed income (G = .3617, K = .3539).

So, considering the above two arguments, we can obviously not conclude that the structure of final demand, and the changes in it between 1959 and 1969, "did not matter."

This can, more convincingly, be shown by comparing the inequality measures associated with each final demand component of the 1959 model with those of experiment 1. This is done in Table IV.4.

We notice that the changes in final demand, for every component, were a cause of increased income inequality. Since simulation experiment 1 holds every model component constant at its 1959 value except exogenous final demand, this increase in inequality is thus due only to changes in the sectoral composition of final demand. 1/

1/ In section III. 3.2 we also noted that for 1969 the household incomes generated by each final demand component were distributed less equally than in 1959. This result was obtained though by comparing the 1959 model solution with the 1969 one, in other words, it showed the effect of changes in all model elements. Experiment 1 here singles out the increase in inequality due to changes in final demand structure only.

Table IV.4. Inequality Associated with Each Final Demand Component, 1959 and Experiment 1

	C_o	G	I	I_{nv}	X	X_s	M
1959 G	.2545	.2399	.2072	.2163	.2276	.2189	.1959
K	.2502	.2291	.2045	.2131	.2228	.2155	.1911
Experiment 1 G	.2680	.2465	.2103	.2288	.2308	.2305	.2049
K	.2636	.2370	.2073	.2245	.2258	.2248	.1984

Source: Table III.8, also see text.

It is worth analyzing these changes in some more detail.

The sectoral change in consumption outside households toward services promoted inequality as the incomes resulting from spending on the services sector are distributed less equally (G = .2850, K = .2805) than the overall income distribution (which has G = .2277, K =.2230). 1/ In addition, the shift was away from retail and wholesale and miscellaneous, two sectors with less inequality in the incomes they generate than services. Mutatis mutandis, the same applies to government spending, as the losing sector (government services) has a low level of inequality (G = .2058, K = .1899), while the gaining sectors (services, finance, insurance and real estate, utilities) all have higher levels of inequality.

Investment saw a reduction in construction, a sector that has very low inequality associated with it (G = .1948, K =.1926), toward manufacturing and wholesale and retail which generate less equally distributed incomes. The same applies to change in inventories and exports where the shift was away from agriculture, which is the sector that generates the most equally distributed incomes (G = .1538, K = .1705).

The increase in inequality associated with imports (had they been produced domestically) is an element that tends to reduce inequality of the distribution of total household incomes, since the latter are generated by final demand including net exports. 2/

1/ These and following figures are quoted from Tables III.6 and IV.2.

2/ I.e. $C_o + G + I + I_{nv} + (X + X_s - M)$.

We may conclude that all the major changes in the structure of final demand (with the exception of those in imports) tend to promote a move toward inequality. The fact that the overall effect as registered by the change in G and K between the 1959 model solution and experiment 1 is so small is due, in part, to the fact that the increase in inequality in the sector which is most important in absolute figures, viz. investment, was fairly small.

It is also due, in part, to the inability of an inequality summary measure to adequately capture changes in distribution when Lorenz-curves intersect. When comparing the equilibrium income distribution of the 1959 model solution with the one of experiment 1, we noticed that the income shares of both the poorest and the richest quintiles were higher in experiment 1, while the shares of the three middle groups fell.

Considering this, and our analysis of the distributional implications of the sectoral shifts in each final demand component, we believe it is warranted to conclude that the registered increase of G and K, while numerically small, is not insignificant and represents a true unequalizing effect of changes in the composition of final demand.

We can be briefer in reporting about the other simulation experiments.

Experiment 2 brings the 1969 input-output structure into the 1959 model. This results in an increase in inequality and a reduction of the total household income multiplier.

Experiment 3 brings the 1969 final demand structure into the experiment 2 model, which, again, increases inequality a little bit.

Experiment 4 shows the impact of the 1969 consumption pattern. This reduces income inequality and also the level of the income multiplier.

Experiment 5 shows again that the 1969 exogenous final demand structure slightly increases overall inequality (the increase, in this case, is only registered by K).

Experiment 6 introduces the 1969 V-matrix which raises inequality substantially (in fact, to a level above the one eventually observed in 1969). This is consistent with the results of the studies that pointed at the tightening of the labor market in Japan in the sixties and its impact on inequality (see section III.1). The distribution of households' share of value added is indeed greatly influenced by labor market conditions.

Experiment 7, which brings the 1969 final demand into experiment 6, strangely enough, reduces inequality this time. This may appear inconsistent with the finding of experiments 1, 3 and 5 discussed above, but actually it is not. To understand this we must look at the $(I - VBc')^{-1}$ VB submatrix, which links final demand to income distribution, and see how it is changed when V_{1959} is replaced with V_{1969}. Table IV.5 compares the inequality summary measures G and K implied by the 1959 submatrix and those implied by the same submatrix in experiment 7.

We notice immediately quite substantial differences in the distribution of incomes generated by each input-output sector. We shall now reanalyze, in the light of these differences, the changes in the pattern of final demand that took place between 1959 and 1969.

Consumption outside households and government spending shifted to services. With the 1959 V-matrix in effect, services generated income quite unequally distributed (second highest G and K); with the 1969 V-matrix in effect, as in experiment 7, the income generated by services is relatively more equally distributed (only fifth highest G and K). Government spending shifted away from

Table IV.5. Inequality Generated by Each Input-Output Sector, 1959 and Experiment 7

		AGR	MIN	MFG	CON	EGW	WRET
1959	G	.1538	.2081	.2265	.1948	.2365	.2681
	K	.1705	.1918	.2229	.1926	.2238	.2608
Exp. 7	G	.2943	.2153	.2774	.2531	.2999	.2724
	K	.2830	.2004	.2630	.2380	.2908	.2591

		FIR	TC	GOV	SERV	OFF	ANAD
1959	G	.3617	.2455	.2058	.2850	.2351	.2126
	K	.3539	.2330	.1899	.2805	.2306	.2106
Exp. 7	G	.3342	.2609	.3084	.2912	.2736	.2295
	K	.3220	.2365	.2973	.2794	.2597	.2241

Source: Table III.6, also see text.

government services. In 1959 this meant a shift away from the sector with the third most equal distribution, consequently it increased inequality. In experiment 7 it meant a shift away from the sector with the second most unequal distribution, consequently it promoted equality.

For the other final demand components, major gaining sectors are wholesale and retail and manufacturing, a major losing sector is agriculture. In 1959 wholesale and retail ranked third in terms of the inequality of incomes it generates, in experiment 7 it ranked eighth. Manufacturing kept its relative position (seventh in 1959, sixth in experiment 7). Agriculture had the least inequality associated with it in 1959, in experiment 7 it ranked fourth in terms of inequality.

Conclusion of all this is that the shifts in final demand between 1959 and 1969 were, with the 1959 V-matrix in effect, shifts from sectors with a relatively equal income distribution to sectors with a relatively unequal distribution. In experiment 7, with the 1969 V-matrix operational, it was the other way around. It is thus clear that the same shift in final demand composition can have either an equalizing or an unequalizing effect depending upon the components of the matrix that links final demand with income distribution.

With respect to what actually happened in Japan between 1959 and 1969, we might say tentatively (since our model does not explicitly incorporate time-paths) that in the early sixties when the 1959 V-matrix applied, the changes in final demand (assuming they took place in a fairly linear trend fashion over the decade) promoted inequality. Later on, when the 1969 V-matrix was more representative of reality, the final demand changes slowed down the rise in inequality brought about by the 1969 level and distribution of households' share of value added.

The results of experiments 8-14 further confirm the above analysis. Every simulation involving the 1959 V-matrix produces a rise in inequality when the 1969 final demand is brought in (compare experiments 1, 3, 5, 9 with, respectively, the 1959 model, experiments 2, 4, 8). In the other cases, where the 1969 V-matrix appears, the 1969 final demand structure reduces inequality (compare experiments 7, 11, 13, and the 1969 model with, respectively, experiments 6, 10, 12, 14). [1/]

It is worthwhile to note that the impact of the changes in the Leontief-inverse and in the consumption pattern also differs, depending on whether the 1959 or 1969 V-matrix is in the model. If we interpret this in a similar way as we did for final demand changes, we may conclude the following.

In the first half of the decade between 1959 and 1969, when the distribution in value added was best represented by V_{1959}, the effect of the changes in input-output structure was to increase inequality. In the second half of the decade, when V_{1969} was more relevant, the opposite was true, although the equalizing effect was small.[2/]

1/ The consistency of this pattern is another reason why we believe that the registered changes in G and K are significant, in spite of the fact that they are small numerically.

2/ This conclusion obtains from the following comparisons. When V_{1959} is operational, compare the 1959 model, experiments 1, 4, 5 with experiments 2, 3, 8, 9: introducing the 1969 Leontief-inverse increases G by about .02. When V_{1969} is operational compare experiment 6, 7, 12, 13 with experiments 10, 11, 14, the 1969 model: introducing the 1969 Leontief-inverse reduces G by about .0025. Source: Tables IV.1 and IV.2.

The effect of changes in the consumption pattern is to reduce inequality, but by more in the first half of the decade. 1/

IV.3 Summary

In this chapter we subjected our model to a series of simulation exercises in an attempt to find out what the relative contribution was of several structural changes that took place between 1959 and 1969 to the trend in the equality of income distribution observed during that period.

Our findings can be summarized as follows.

1) The changes in the level and the distribution of households' share of value added between 1959 and 1969 contributed to increasing inequality.

2) The changes in input-output structure reduced household incomes being generated. In the first half of the decade 1959-1969, they tended to increase total inequality, in the second half their effect was mildly equalizing.

3) The changes in consumption patterns had an equalizing effect on total income distribution, especially in the first half of the decade.

4) Substantial shifts occurred in the sectoral composition of each final demand component between 1959 and 1969. In the beginning of the decade this tended to increase inequality. In the later years, when changes in the level and distribution of households' share of value added increased inequality, the changes in the structure of final demand served to somewhat slow down that trend. 2/

1/ This conclusion obtains from the following comparisons. When V_{1959} is operational, compare the 1959 model, experiments 1, 2, 3 with experiments 4, 5, 8, 9: introducing the 1969 consumption pattern reduces G by about .015. When V_{1969} is operational, compare experiments 6, 7, 10, 11 with experiments 12, 13, 14, the 1969 model: introducing the 1969 consumption pattern reduces G by about .004. Source: Tables IV.1 and IV.2.

2/ The same factor, viz. changes in the level and distribution of households' share of value added, is the explanation of why the distributional impact of changes in input-output structure and in consumption patterns was not constant over the entire decade.

CHAPTER V

SUMMARY AND CONCLUSION

This paper investigated the impact on income distribution of changes in the structure of final demand that occur in the process of economic growth.

Theoretical and empirical studies about the relationship between distribution and growth were briefly surveyed in Chapter I. One of the more recent and promising lines of research attempts to integrate the theory of distribution and growth. The key to that attempt is the study of the interdependence between the growth of income in one socio-economic group and the growth of income in another. Socio-economic groups are "linked" together via capital ownership (output produced by capital owned by one group flows in part to other groups in the form of wages) and via consumption demand (the production of the consumption goods for one group results in the payment of wages going in part to other groups). In this study we focused on the linkages via consumption demand. The explanation of the changes in linkages between socio-economic groups over time, and the tying of this explanation to the analysis of overall changes in income distribution, is the major innovative contribution of this study.

The model we used for the analysis was presented in Chapter II. It is a conventional input-output model augmented with a distribution matrix and closed for the household sector which is disaggregated according to one or more socio-economic criteria. The closure of the model effectuates the inclusion of indirect (or feedback) effects from household spending on to production, income generation, and back on to spending. The model yields the equilibrium level and distribution of household income in a manner fully consistent with the production structure of the economy.

Exogenous final demand is broken down into its components and full account is taken of the sectoral composition of each component. This makes possible the derivation of the amount of household income and its distribution generated by each final demand component separately. This form of "distribution-accounting" yields more information about the factors contributing to an observed overall distribution than conventional decomposition of inequality summary measures.

As a result of the explicit recognition of socio-economic groups, the solution of the model contains a sub-matrix which conveniently summarizes the demand linkages between socio-economic groups (it was therefore labelled "linkages matrix") and which yields useful interpretations regarding different levels of inequality which have relevance for policy.

Specifically, it was shown that one can derive from the linkages matrix a lower bound level of inequality which corresponds to the minimum inequality feasible in equilibrium, consistent with certain structural features of the economy. This level, which was labelled "endogenous inequality" can, in the long run, only be obtained by continuous exogenous redistributive efforts. The comparison of endogenous with actual inequality provides useful indications about the appropriate policy tools for affecting the existing distribution.

The linkages matrix also implies a steady state level of inequality which would maintain itself indefinitely if no structural changes occur. We argued that an international comparison of actual, endogenous and steady state levels of inequality would yield valuable information about the causes and underlying mechanisms of distributional patterns and be much more relevant than the customary comparisons of actual levels only.

In Chapter III we presented an empirical application of the model to Japan during the time period 1959-1969. During that decade the Japanese economy grew rapidly and was not disturbed by any major external shocks.

The inequality of the Japanese income distribution increased in the fifties. In the early sixties the trend was reversed, and inequality fell. From the late sixties on though an upward trend of the Gini-coefficient was noted again. The studies that tried to explain these trends focused on demographic factors and changes in the labor market. Demand received little or no attention. The application of a demand oriented model to Japan can therefore yield some new insights into the process of a changing income distribution in that country.

As to the structure of the linkages matrix, we found that higher income groups had a higher within-group multiplier, i.e. their spending generated more income for themselves than was the case for the lower income groups. This was in spite of the fact that, on a per unit basis, income earned and spent by higher income groups generated less total household income than income originally earned by or transferred to poor groups. The web of linkages between income groups changed between 1959 and 1969 to favor the middle income classes.

It was found that during the period under consideration, the actual level of inequality was not much above its steady state level. The endogenous level, however, was substantially below the steady state level.

Since endogenous and steady state inequality decreased between 1959 and 1969, but the distribution of total household income became less equal, the explanation needs to be found in changes in the composition of final demand.

It was found that final demand in different sectors is associated with widely different levels of inequality of the incomes generated. In 1959, high levels of income inequality were generated by final demand in the finance, insurance and real estate sector, and in services. The lowest level of inequality stemmed from final demand in agriculture and construction. In 1969, finance, insurance and real estate, the government sector, and the utilities sector generated the least equally distributed income. Construction and mining were the sectors yielding the lowest levels of inequality.

Of all exogenous final demand components, investment generated the most equal income distribution because of the importance of the construction sector for investment. Since investment was the largest component of exogenous final demand, it played a major role in explaining the low level of inequality in the distribution of household incomes in 1959.

The increased importance of investment as a source of income in 1969, combined with a less equal distribution of that income (traceable to sectoral shifts within investment away from construction), explained in part the observed increase in income inequality between 1959 and 1969. At the same time it helped to check that increase, which indeed would have been larger had government spending (the second largest component of exogenous final demand) kept its 1959 relative importance, since the increase in inequality was greater in the incomes derived from government spending than in those derived from investment.

Chapter IV presented the results of a series of simulation exercises which attempted to discover the contribution of several structural changes that took place between 1959 and 1969 to observed trends in inequality.

The changes in the level and distribution of households' share of value added between 1959 and 1969 contributed to increasing inequality. This finding is consistent with studies of the Japanese labor market over that period.

In the first part of the 1959-1969 decade changes in input-output structure and changes in the composition of exogenous final demand tended to increase total inequality. The changes in consumption pattern had an equalizing effect. In the last part of the decade, when changes in the level and distribution of households' share of value added had increased inequality, the changes in input-output structure, in consumption and exogenous final demand all cooperated in slowing down the increasing trend in inequality. As said above, the role of investment stood out in this process.

Finally, a word needs to be said about the limitations of the model used. These limitations are by implication suggestions for future research. First, the model is not able to trace the effects of changes in relative prices. This could be remedied by the incorporation of, for example, a linear expenditure system which would allow for price effects on expenditure patterns. Second, the value added shares are fixed and given, which assumes an exogenously determined and fixed distribution of ownership of resources. Adding a labor market model would, in part, solve this problem. It seems unlikely that it is feasible to construct a model of capital goods distribution with the present state of data availability. Third, capacity constraints could be acknowledged by adding an investment matrix to the model. Fourth, the study of the linkages between socio-economic groups could usefully be done, data permitting, along other dimensions than income, for example, occupation. Finally, in the spirit of semi-input-output models, the endogenous variable in the model (the level and distribution of household income) could be made into a pre-specified target and the model then run in reverse to determine the required level of certain policy instruments such as exports, income transfers, etc. that are compatible with the target.

We believe that the methodology and the empirical application and findings of this paper present some new insights into the process of income distribution which could not have been derived from distributional studies using alternative approaches, and which might warrant the undertaking of similar studies for other countries.

APPENDIX ONE: Correspondence Between NFIE and Input-Output Categories

In order to make the NFIE categories comparable to those of the input-output table some subdividing and re-arranging of NFIE categories was necessary, as well as a number of conceptual adjustments. The scheme of correspondence is presented in Table A1-1.

The categories Food and Beverages, Other Fuel and Light, Clothing, Medical Care, Toilet Care, Private Transportation, Reading and Recreation needed to be subdivided and assigned to more than one input-output category. In these cases the more detailed breakdown table of household spending per commodity (59 NFIE, Vol. III, Table 1-1, p. 31-180; 69 FIES, Table 15, p. 288-373) was used to obtain the percentage assignment of these NFIE categories to the relevant input-output categories. A distinction was made (i.e. different percentages were computed) for workers' households and for other households in 1969.

The sum of all these regrouped expenditure categories was labeled "Adjusted Expenditures." This is equal to "Living Expenditures" in NFIE minus Remittances, which cannot be assigned to any input-output category, as it is simply a money transfer from one household to another and does not generate any direct productive activity (except perhaps for a postal or banking service used to make the remittance).

There are also some conceptual differences between the categories of a household spending survey and those of an input-output table.

The major problem has to do with credit purchases. The NFIE lists the total price (i.e. including interest payments and service fees) under the relevant expenditure category. A balancing entry is made under "Credit purchase," a subcategory of "Receipts other than income." The monthly payments do not enter in the expenditure breakdown at all. They are listed as "Credit purchase payments," a subcategory of "Disbursements other than expenditures" (1959 NFIE, Vol. I, p. 18; 1969 NFIE, Vol. I, p. 44).

Strictly speaking, a credit purchase of, for example, a radio (which is listed in a household survey at consumers' price) ought to go to three input-output categories: "manufacturing" for the producers' sale price of the item, "wholesale and retail" for the mark-up margins of the distributors, "finance" for interest payments and service fees of the financial institution.

Since the input-output table we use is based on producers' prices, all trade margins are recorded on the row of the Wholesale and Retail sector (Link Input-Output Tables, Explanatory Report, 1975, p. 110). We have not attempted though to further separate off the Finance sector fees and interests, since, with the data at hand, this would necessitate the making of some arbitrary and guesswork assumptions about the relative share of these fees in total price.

A related problem pertains to insurance premium payments. In the cases of fire and indemnity insurance, the premium is a payment for deferring risks to an insurance company and as such fully assignable to the input-output category "Insurance." Life insurance, on the other hand, is often a mixed form of insurance and savings, viz. in the case when the policyholder receives a certain amount of money at the end of the contract period when he is still alive. The NFIE lists life insurance premiums as a "Disbursement other than expenditure" which would be appropriate if the majority of the life insurance policies would be of the mixed variety. As we have no information on the total percentage of premium payments that is savings and the percentage that is payment for having the risk of death "covered," we chose not to assign this category to the input-output sector "Insurance."

A last problem has to do with "Social Expenses," which is really a use classification (it includes all expenses to visitors--1959 NFIE, p. 18, p. 440) intermingled in the commodity classification (Food, Beverages, Clothing...) used by NFIE for all other expenditures. While most of this can be assigned to services, not all of it can. But again, for simplicity and lack of any basis for splitting, we assigned all of it to the input-output category "Services."

Table A1-1: Correspondence Between the NFIE and the Input-Output Categories

Input-Output Category	NFIE Category
1. Agriculture (AGR)	rice, barley and other cereals, fresh fish, fresh vegetables, fruits
2. Minerals (MIN)	other fuel and light (except those classified as manufacturing)
3. Manufacturing (MFG)	all food and beverages (except those classified as agriculture, and except food prepared outside household)
	furniture and utensils
	other fuel and light (petroleum, propane gas, other fuel, ice)
	clothing (except tailoring fees, laundry, clothing rent)
	medical care (all drugs and hygienic materials)
	toilet care (soaps, cosmetics, insecticides, toilet tissues, tooth brushes)
	private transportation (except car insurance)
	stationary
	reading and recreation (newspapers, magazines, books, recreational goods)
	tobacco
4. Construction (CON)	repairs and improvements
5. Electricity, gas & water (EGW)	water, electricity and gas
6. Wholesale & retail trade (WRET)	

Table A1-1: (continued) Correspondence Between the NFIE and the Input-Output Categories

Input-Output Category	NFIE Category
7. Finance, insurance, real estate (FIR)	rent private transportation (car insurance) indemnity and fire insurance
8. Transportation & communication (TC)	transportation and communication
9. Government service (GOV)	
10. Services (SERV)	food prepared outside household clothing (other than classified as manufacturing) medical care (other than classified as manufacturing) toilet care (other than classified manufacturing) reading and recreation (other than classified as manufacturing) education social expenses, obligation fees, other miscellaneous
11. Office supplies and packing materials (OFF)	
12. Activities not adequately described (ANAD)	

Source: 1959 NFIE, Vol. I, Appendix 1, p. 425-440.
1960-65-70 Link Input-Output Tables, p. 8-18.

REFERENCES

Adelman, I. and Morris, C., 1971. An Anatomy of Patterns of Income Distribution in Developing Countries, Stanford: Stanford University Press.

Adelman, I. and Morris, C., 1973. Economic Growth and Social Equity in Developing Countries, Stanford: Stanford University Press.

Adelman, I. and Robinson, S., 1976. Income Distribution, Import Substitution and Growth Strategies in a Developing Country, Research Program in Development Studies, Woodrow Wilson School, Princeton University, Discussion Paper #68.

Ahluwalia, M., 1976a. Income Distribution and Development: Some Stylized Facts, American Economic Review, May 1976, p. 128-132.

Ahluwalia, M., 1976b. Inequality, Poverty and Development, Journal of Development Economics, December 1976, p. 307-342.

Ahluwalia, M., and Chenery, H., 1975. A Model of Distribution and Growth in: Chenery et al., Redistribution with Growth, London: Oxford University Press, 1975, p. 209-235.

Arrow, K. and Hoffenberg, M., 1959. A Time Series Analysis of Interindustry Demands, Amsterdam: North-Holland Publishing Company.

Ballentine, J. and Soligo, R., 1974. Consumption and Earnings Patterns and Income Distribution, mimeo, Program of Development Studies, Rice University, Houston, Texas.

Blitzer, C., Clark, P. and Taylor, L. (eds.), 1975. Economy-wide Models and Development Planning, New York: Oxford University Press for the World Bank.

Boltho, A., 1975. Japan - An Economic Survey 1953-1973, Economies of the World, New York: Oxford University Press.

Chenery, H. and Clark, P., 1959. Interindustry Economics, New York: John Wiley and Sons.

Chenery, H. and Syrquin, M., 1975. Patterns of Development 1950-1970, London: Oxford University Press.

Chenery, H. et al., 1975. Redistribution with Growth, London: Oxford University Press.

Chinn, D., 1972. Effects of Income Redistribution on Economic Growth Constraints: Evidence from the Republic of Korea, Economic Bulletin for Asia and the Far East, 1972, p. 61-76.

Chinn, D., 1977. Distributional Equality and Economic Growth: The Case of Taiwan, Economic Development and Cultural Change, October 1977, p. 65-80.

Clark, P., 1975. Intersectoral Consistency and Macroeconomic Planning, in: Blitzer, C., Clark, P., and Taylor, L. (eds.), Economy-wide Models and Development Planning, New York: Oxford University Press for the World Bank, 1975, p. 129-154.

Cline, W.R., 1972. Potential Effects of Income Redistribution on Economic Growth: Latin American Cases, New York: Praeger Publishers.

Cole, R. and Tominaga, K., 1976. Japan's Changing Occupational Structure and its Significance, in: Patrick, H. (ed.), Japanese Industrialization and its Social Consequences, Berkeley: University of California Press, 1976, p. 53-96.

Economic Planning Agency, 1972. Economic Survey of Japan 1971-72, Tokyo: Economic Planning Agency, Government of Japan.

Figueroa, A., 1975. Income Distribution, Demand Structure and Employment: The Case of Peru, Journal of Development Studies, January 1975, p. 20-31.

Foxley, A., 1976. Redistribution of Consumption Effects on Production and Employment, Journal of Development Studies, April 1976, p. 171-190.

Goldberger, A., 1964. Econometric Theory, New York: John Wiley and Sons.

Grootaert, C., 1978. Patterns of Final Demand and Income Distribution: The Case of Japan, Ph.D. Dissertation, University of Hawaii.

Income Distribution, Employment and Economic Development in Southeast and East Asia, Papers and Proceedings of the Seminar sponsored jointly by the Japan Economic Research Center (J.E.R .C) and the Council for Asian Manpower Studies (C.A.M.S), December 16-20, 1974, Tokoyo, Manila, J.E.R.C.-C.A.M.S, 1975.

Ishizaki, T., 1967. The Income Distribution in Japan, The Developing Economies, No. 2, 1967, p. 351-370.

Kravis, I.B., 1960. International Differences in the Distribution of Income, Review of Economics and Statistics, November 1960, p. 408-416.

Kuznets, S., 1955. Economic Growth and Income Inequality, American Economic Review, March 1955, p. 1-28.

Kuznets, S., 1963. Quantitative Aspects of the Economic Growth of Nations: VIII. Distribution of Income by Size, Economic Development and Cultural Change, January 1963, p. 1-80.

Kuznets, S., 1975. Summary, in: Income Distribution, Employment and Economic Development in Southeast and East Asia, Papers and Proceedings of the Seminar sponsored jointly by the Japan Economic Research Center (J.E.R.C.) and the Council for Asian Manpower Studies (C.A.M.S.), December 16-20, 1974, Tokyo, Manila, J.E.R.C.-C.A.M.S., 1975, p. 789-791.

Lal, D., 1976. Distribution and Development: A Review Article, World Development, September 1976, p. 725-737.

Land, J.W. and Soligo, R., 1971. Income Distribution, Employment and Growth in Labor Redundant Economies, Discussion Paper #9, Program of Development Studies, Rice University, Houston, Texas.

Leontief, W. et al., 1953. Studies in the Structure of the American Economy, New York: Oxford University Press.

Looney, R.E., 1975. Income Distribution and Economic Growth in Semi-Industrialized Countries: A Comparative Study of Iran, Mexico, Brazil and South Korea, New York: Praeger Publishers.

Lopes, F.L., 1973. Inequality and Growth: A Programming Model with Applications to Brazil, Cambridge, Mass.: Development Research Group, Center for International Affairs, Harvard University.

Minami, R., 1973. The Turning Point in Economic Development: Japan's Experience, Tokyo: Kinokuniya Bookstore.

Miyazawa, K., 1976. Input-Output Analysis and the Structure of Income Distribution, New York: Springer Verlag.

Mizoguchi, T., 1975a. Size Distribution of Household Income in Postwar Japan, in: Income Distribution, Employment and Economic Development in Southeast and East Asia, Papers and Proceedings of the Seminar sponsored jointly by the J.E.R.C. and the C.A.M.S., December 16-20, 1974, Tokyo, Manila, J.E.R.C.-C.A.M.S, 1975, p. 473-515.

Mizoguchi, T., 1975b. Income and Assets Distribution in Japan and South Korea, Manila: Council for Asian Manpower Studies, Discussion Paper #75-07.

Mizoguchi, T., Takayama, N., and Terasaki, Y., 1977. Over-Time Changes of Size Distribution of Household Income Under the Rapid Economic Growth: The Japanese Experience, Institute of Economic Research, Hitotsubashi University, Tokyo, Working Paper J-6.

Morley, S.A. and Williamson, J., 1974. Demand, Distribution and Employment: The Case of Brazil, Economic Development and Cultural Change, October 1974, p. 33-60.

Mouer, R., 1973, 1974. Income Distribution in Japan: An Examination of FIES Data 1963-71, Keio Economic Studies #1, 1973; #1, 2, 1974.

Ono, A. and Watanabe, T., 1976. Changes in Income Inequality in the Japanese Economy, in: Patrick, H. (ed.), Japanese Industrialization and its Social Consequences, Berkeley: University of California Press, 1976, p. 363-390.

Oshima, H.T., 1962. The International Comparison of Size Distribution of Family Incomes with Special Reference to Asia, Review of Economics and Statistics, November 1962, p. 439-445.

Oshima, H.T., 1970. Income Inequality and Economic Growth - the Postwar Experience of Asian Countries, Malayan Economic Review, October 1970, p. 7-41.

Patrick, H. (ed.), 1976. Japanese Industrialization and its Social Consequences, Berkeley: University of California Press.

Patrick, H. and Rosovsky, H. (eds.), 1976a. Asia's New Giant - How the Japanese Economy Works, Washington, D.C.: Brookings Institution.

Patrick, H. and Rosovsky, H., 1976b. Japan's Economic Performance: An Overview, in: Patrick H. and Rosovsky, H. (eds.), Asia's New Giant - How the Japanese Economy Works, Washington, D.C.: Brookings Institution, 1976.

Paukert, F., 1973. Income Distribution at Different Levels of Development: A Survey of Evidence, International Labour Review, August-September 1973, p. 97-125.

Paukert, F., Skolka, J. and Maton, J., 1975. Redistribution of Income, Patterns of Consumption and Employment - A Case Study for the Philippines, in: Income Distribution, Employment and Economic Development in Southeast and East Asia, Papers and Proceedings of the Seminar sponsored jointly by the Japan Economic Research Center (J.E.R.C.) and the Council for Asian Manpower Studies (C.A.M.S), December 16-20, 1974, Tokyo, Manila, J.E.R.C-C.A.M.S., 1975, p. 197-231.

Ranis, G., 1971. Employment and Income Distribution in Developing Countries, I.B.R.D., Economic Staff Working Paper #115.

Ranis, G., 1975. LDC Employment and Growth: A Synthesis of Economic Growth Center Research, Economic Growth Center, Yale University, Center Discussion Paper #231.

Sawyer, M., 1976. Income Distribution in O.E.C.D. Countries, O.E.C.D., Paris.

Soligo, R., 1973. Factor Intensity of Consumption Patterns, Income Distribution and Employment Growth in Pakistan, Program of Development Studies Paper #44, Rice University, Houston, Texas.

Soligo, R. and Land, J.W., 1972. Models of Development Incorporating Distribution Aspects, Program of Development Studies Paper #22, Rice University, Houston, Texas.

Stewart, J., 1978, Potential Effects of Income Redistribution on Economic Growth: An Expanded Estimating Procedure Applied to Mexico, Economic Development and Cultural Change, April 1978, p. 467-485.

Sunman, T., 1973. Short-Run Effects of Income Distribution on Some Macro-Economic Variables: The Case of Turkey, Program of Development Studies Paper #46, Rice University, Houston, Texas.

Takahashi, C., 1959. Dynamic Changes of Income and its Distribution in Japan, Tokyo: Kinokuniya Bookstore.

Turnham, D.J. and Hawkins, E.K., 1973. The Employment Problem and World Bank Activities, I.B.R.D., Staff Working Paper #148.

Wada, R.O., 1975. Impact of Economic Growth on the Size Distribution of Income: The Postwar Experience of Japan, Ph.D. dissertation, University of Hawaii.

Weisskoff, R., 1976. Income Distribution and Export Promotion in Puerto Rico, Yale Economic Growth Center, Yale University, Center Paper #245.

Wonnacott, P., 1974. Macroeconomics, Homewood, Illinois: R. Irwin Inc.

Data Sources

Japan, Administrative Management Agency, 1960-1965-1970 Link Input-Output Tables, Explanatory Report, Data Report (1), Tokyo, Administrative Management Agency, 1974-75.

Japan, Economic Planning Agency, Annual Report on National Income Statistics, 1970, 1975. Tokyo, Economic Planning Agency, 1970, 1975.

Japan, Office of the Prime Minister, Annual Report on the Family Income and Expenditure Survey, 1959, 1969. Tokyo, Bureau of Statistics, Office of the Prime Minister, 1960, 1971.

Japan, Office of the Prime Minister, Japan Statistical Yearbook, 1960, 1970. Tokyo, Bureau of Statistics, Office of the Prime Minister, 1960, 1971.

Japan, Office of the Prime Minister, National Survey of Family Income and Expenditures, 1959, 1969. Tokyo, Bureau of Statistics, Office of the Prime Minister, 1961, 1971.

Vol. 157: Optimization and Operations Research. Proceedings 1977. Edited by R. Henn, B. Korte, and W. Oettli. VI, 270 pages. 1978.

Vol. 158: L. J. Cherene, Set Valued Dynamical Systems and Economic Flow. VIII, 83 pages. 1978.

Vol. 159: Some Aspects of the Foundations of General Equilibrium Theory: The Posthumous Papers of Peter J. Kalman. Edited by J. Green. VI, 167 pages. 1978.

Vol. 160: Integer Programming and Related Areas. A Classified Bibliography. Edited by D. Hausmann. XIV, 314 pages. 1978.

Vol. 161: M. J. Beckmann, Rank in Organizations. VIII, 164 pages. 1978.

Vol. 162: Recent Developments in Variable Structure Systems, Economics and Biology. Proceedings 1977. Edited by R. R. Mohler and A. Ruberti. VI, 326 pages. 1978.

Vol. 163: G. Fandel, Optimale Entscheidungen in Organisationen. VI, 143 Seiten. 1979.

Vol. 164: C. L. Hwang and A. S. M. Masud, Multiple Objective Decision Making – Methods and Applications. A State-of-the-Art Survey. XII, 351 pages. 1979.

Vol. 165: A. Maravall, Identification in Dynamic Shock-Error Models. VIII, 158 pages. 1979.

Vol. 166: R. Cuninghame-Green, Minimax Algebra. XI, 258 pages. 1979.

Vol. 167: M. Faber, Introduction to Modern Austrian Capital Theory. X, 196 pages. 1979.

Vol. 168: Convex Analysis and Mathematical Economics. Proceedings 1978. Edited by J. Kriens. V, 136 pages. 1979.

Vol. 169: A. Rapoport et al., Coalition Formation by Sophisticated Players. VII, 170 pages. 1979.

Vol. 170: A. E. Roth, Axiomatic Models of Bargaining. V, 121 pages. 1979.

Vol. 171: G. F. Newell, Approximate Behavior of Tandem Queues. XI, 410 pages. 1979.

Vol. 172: K. Neumann and U. Steinhardt, GERT Networks and the Time-Oriented Evaluation of Projects. 268 pages. 1979.

Vol. 173: S. Erlander, Optimal Spatial Interaction and the Gravity Model. VII, 107 pages. 1980.

Vol. 174: Extremal Methods and Systems Analysis. Edited by A. V. Fiacco and K. O. Kortanek. XI, 545 pages. 1980.

Vol. 175: S. K. Srinivasan and R. Subramanian, Probabilistic Analysis of Redundant Systems. VII, 356 pages. 1980.

Vol. 176: R. Färe, Laws of Diminishing Returns. VIII, 97 pages. 1980.

Vol. 177: Multiple Criteria Decision Making-Theory and Application. Proceedings, 1979. Edited by G. Fandel and T. Gal. XVI, 570 pages. 1980.

Vol. 178: M. N. Bhattacharyya, Comparison of Box-Jenkins and Bonn Monetary Model Prediction Performance. VII, 146 pages. 1980.

Vol. 179: Recent Results in Stochastic Programming. Proceedings, 1979. Edited by P. Kall and A. Prékopa. IX, 237 pages. 1980.

Vol. 180: J. F. Brotchie, J. W. Dickey and R. Sharpe, TOPAZ – General Planning Technique and its Applications at the Regional, Urban, and Facility Planning Levels. VII, 356 pages. 1980.

Vol. 181: H. D. Sherali and C. M. Shetty, Optimization with Disjunctive Constraints. VIII, 156 pages. 1980.

Vol. 182: J. Wolters, Stochastic Dynamic Properties of Linear Econometric Models. VIII, 154 pages. 1980.

Vol. 183: K. Schittkowski, Nonlinear Programming Codes. VIII, 242 pages. 1980.

Vol. 184: R. E. Burkard and U. Derigs, Assignment and Matching Problems: Solution Methods with FORTRAN-Programs. VIII, 148 pages. 1980.

Vol. 185: C. C. von Weizsäcker, Barriers to Entry. VI, 220 pages. 1980.

Vol. 186: Ch.-L. Hwang and K. Yoon, Multiple Attribute Decision Making – Methods and Applications. A State-of-the-Art-Survey. XI, 259 pages. 1981.

Vol. 187: W. Hock, K. Schittkowski, Test Examples for Nonlinear Programming Codes. V. 178 pages. 1981.

Vol. 188: D. Bös, Economic Theory of Public Enterprise. VII, 142 pages. 1981.

Vol. 189: A. P. Lüthi, Messung wirtschaftlicher Ungleichheit. IX, 287 pages. 1981.

Vol. 190: J. N. Morse, Organizations: Multiple Agents with Multiple Criteria. Proceedings, 1980. VI, 509 pages. 1981.

Vol. 191: H. R. Sneessens, Theory and Estimation of Macroeconomic Rationing Models. VII, 138 pages. 1981.

Vol. 192: H. J. Bierens: Robust Methods and Asymptotic Theory in Nonlinear Econometrics. IX, 198 pages. 1981.

Vol. 193: J. K. Sengupta, Optimal Decisions under Uncertainty. VII, 156 pages. 1981.

Vol. 194: R. W. Shephard, Cost and Production Functions. XI, 104 pages. 1981.

Vol. 195: H. W. Ursprung, Die elementare Katastrophentheorie. Eine Darstellung aus der Sicht der Ökonomie. VII, 332 pages. 1982.

Vol. 196: M. Nermuth, Information Structures in Economics. VIII, 236 pages. 1982.

Vol. 197: Integer Programming and Related Areas. A Classified Bibliography. 1978 – 1981. Edited by R. von Randow. XIV, 338 pages. 1982.

Vol. 198: P. Zweifel, Ein ökonomisches Modell des Arztverhaltens. XIX, 392 Seiten. 1982.

Vol. 199: Evaluating Mathematical Programming Techniques. Proceedings, 1981. Edited by J.M. Mulvey. XI, 379 pages. 1982.

Vol. 201: P. M. C. de Boer, Price Effects in Input-Output-Relations: A Theoretical and Empirical Study for the Netherlands 1949–1967. X, 140 pages. 1982.

Vol. 202: U. Witt, J. Perske, SMS – A Program Package for Simulation and Gaming of Stochastic Market Processes and Learning Behavior. VII, 266 pages. 1982.

Vol. 203: Compilation of Input-Output Tables. Proceedings, 1981. Edited by J. V. Skolka. VII, 307 pages. 1982.

Vol. 204: K.C. Mosler, Entscheidungsregeln bei Risiko: Multivariate stochastische Dominanz. VII, 172 Seiten. 1982.

Vol. 205: R. Ramanathan, Introduction to the Theory of Economic Growth. IX, 347 pages. 1982.

Vol. 206: M.H. Karwan, V. Lotfi, J. Telgen, and S. Zionts, Redundancy in Mathematical Programming. VII, 286 pages. 1983.

Vol. 207: Y. Fujimori, Modern Analysis of Value Theory. X, 165 pages. 1982.

Vol. 208: Econometric Decision Models. Proceedings, 1981. Edited by J. Gruber. VI, 364 pages. 1983.

Vol. 209: Essays and Surveys on Multiple Criteria Decision Making. Proceedings, 1982. Edited by P. Hansen. VII, 441 pages. 1983.

Vol. 210: Technology, Organization and Economic Structure. Edited by R. Sato and M.J. Beckmann. VIII, 195 pages. 1983.

Vol. 211: P. van den Heuvel, The Stability of a Macroeconomic System with Quantity Constraints. VII, 169 pages. 1983.

Vol. 212: R. Sato and T. Nôno, Invariance Principles and the Structure of Technology. V, 94 pages. 1983.

Vol. 213: Aspiration Levels in Bargaining and Economic Decision Making. Proceedings, 1982. Edited by R. Tietz. VIII, 406 pages. 1983.

Vol. 214: M. Faber, H. Niemes und G. Stephan, Entropie, Umweltschutz und Rohstoffverbrauch. IX, 181 Seiten. 1983.

Vol. 215: Semi-Infinite Programming and Applications. Proceedings, 1981. Edited by A. V. Fiacco and K. O. Kortanek. XI, 322 pages. 1983.

Vol. 216: H. H. Müller, Fiscal Policies in a General Equilibrium Model with Persistent Unemployment. VI, 92 pages. 1983.

Vol. 217: Ch. Grootaert, The Relation Between Final Demand and Income Distribution. XIV, 105 pages. 1983.

GPSR Compliance

The European Union's (EU) General Product Safety Regulation (GPSR) is a set of rules that requires consumer products to be safe and our obligations to ensure this.

If you have any concerns about our products, you can contact us on ProductSafety@springernature.com

In case Publisher is established outside the EU, the EU authorized representative is:

Springer Nature Customer Service Center GmbH
Europaplatz 3
69115 Heidelberg, Germany

Batch number: 10844827

Printed by Printforce, the Netherlands